Historical Geography and Geosciences

This book series serves as a broad platform for contributions in the field of Historical Geography and related Geoscience areas. The series welcomes proposals on the history and dynamics of place and space and their influence on past, present and future geographies including historical GIS, cartography and mapping, climatology, climate history, meteorology and atmospheric sciences, environmental geography, hydrology, geology, oceanography, water management, instrumentation, geographical traditions, historical geography of urban areas, settlements and landscapes, historical regional studies, history of geography and historic geographers and geoscientists among other topically related areas and other interdisciplinary approaches. Contributions on past (extreme) weather events or natural disasters including regional and global reanalysis studies also fit into the series.

Publishing a broad portfolio of peer-reviewed scientific books Historical Geography and Geosciences contains research monographs, edited volumes, advanced and undergraduate level textbooks, as well as conference proceedings. This series appeals to scientists, practitioners and students in the fields of geography and history as well as related disciplines, with exceptional titles that are attractive to a popular science audience.

If you are interested in contributing to this book series, please contact the Publisher.

Suresh Bhalai

The History of Mining and Geological Surveys in Jamaica

Springer

Suresh Bhalai
Mines and Geology Division
Ministry of Agriculture, Fisheries
and Mining
Kingston, Jamaica

ISSN 2520-1379 ISSN 2520-1387 (electronic)
Historical Geography and Geosciences
ISBN 978-3-031-42603-2 ISBN 978-3-031-42604-9 (eBook)
https://doi.org/10.1007/978-3-031-42604-9

This Springer imprint is published by the registered company Springer Nature Switzerland AG
The registered company address is: Gewerbestrasse 11, 6330 Cham, Switzerland

Paper in this product is recyclable.

And it has been a pleasant experience to perform it [geological surveys] in a country whose scenic beauty, only those who have been able to, as I have, to explore it thoroughly, can fully appreciate.

Charles Alfred Matley, Government Geologist (October 1921–April 1924). Written at sea on his departure, on the SS Bayano, 25 April 1924, Kingston, Jamaica to Avonmouth, England (The Gleaner, 25 July 1924: 14)

Foreword

The History of Mining and Geological Surveys in Jamaica has sought to encompass the whole history of mining in Jamaica as it can be ascertained from our earliest knowledge of the activities of the indigenous Taino inhabitants, to the complexities of mining processes, together with the associated institutional, legal, and political ramifications unfolding at the present day. This book diverges from some previous (academic) accounts of the development of the knowledge and descriptions of the various types and stratigraphic relationships of the rocks and minerals of the island, in favour of introducing the persons who had to make decisions on the best way of acquiring such knowledge.

A major theme of the author throughout the book has been the detailing of the to-and-fro activities of impoverished governments, reluctant to take on responsibility for providing the resources for properly designed surveys, versus recognizing the needs of prospectors and farmers to have adequate information and infrastructure on hand for pursuing their activities. He emphasises how such needs led to the birth and formation of initially short-lived national geological surveys in the 19th and early 20th Centuries, leading to the Island's present geological survey organisation, and the associated specialist departments, encouraging, guiding, and controlling the applied aspects of natural resources utilisation, such as mining, quarrying and water supplies. This theme is enhanced by more personal, fascinating accounts of the activities of some of the major players involved, both those hired to carry out the practical survey operations and those government officials and parliamentarians who had to juggle financial resources against the needs of the people. This book is to be commended for fleshing out the behind-the-scenes decisions made, sometimes without adequate scientific advice, so that knowledge of Jamaica's geological resources could be improved. It also brings right up to date some of the recent changes in ownership of companies involved in the bauxite industry, with their political consequences.

Kingston, Jamaica — Edward Robinson

Preface

The story of the development of mining and geological research in a country is oftentimes afforded minimal attention in literature. For Jamaica, the case is somewhat better, as there are abundant recordings. These, however, are quite brief to miniscule, scattered over many titles and span several hundreds of years of publications since the Spanish occupation of the Island about 500 years ago. *The History of Mining and Geological Surveys in Jamaica* provides a broad synopsis of the sequences of events and the people that have led to the development of Jamaica's geological assets. This account is presented in generally simplified language to enable the reader to appreciate the construct of the story.

Chapter 1 recognizes the first inhabitants of Jamaica, the Taino people, who were the first to search out and use the mineral resources in their modest way of life. The Spaniards and Englishmen who came later were more concerned with precious and base metal mining, but with no great fortunes. Chapters 2 and 3 elaborate on the role of the English in expanding geological research and undertaking mining ventures. The English were responsible for the execution of private and publicly funded geological surveys that led to increasing knowledge of the Island. Geological surveys were undertaken to investigate mineral and water resources, and support infrastructure development. The greatest boom in the development of Jamaica's mineral industry came with the discovery of bauxite and the changes it brought to the Mining Sector which is described in Chap. 4. The birth of the bauxite industry interestingly, occurred, though independently, at that time when the British Government sought to 'wean' its colonies by promoting their economic development and independence. This led to many modern geological surveys being established in the British Empire, and Jamaica was fortunate to be among these nations as explained in Chap. 5. With over 70 years of bauxite mining and the resulting dwindling reserves, Jamaica now seeks to strategically reposition its Mining Sector, and a discussion is presented in Chap. 6.

The reader will recognize that minerals and mining can be a route towards economic development. They will also notice that government-commissioned geological surveys are a necessity and are oftentimes market-driven rather than based on scientific curiosity. Geological surveys form the basis for the development of productive and profitable mines. Therefore, with sound leadership and management, a nation can optimize the tangible and intangible benefits from their natural resources. This book is prepared not only for scholars and earth science practitioners but is so arranged for anyone wishing to explore this little-known segment of Jamaica's history. Presented herein is *The History of Mining and Geological Surveys in Jamaica* as a wholesome volume.

Kingston, Jamaica Suresh Bhalai

Acknowledgements

Gratitude is expressed to all who contributed to making this publication possible. Foremost I recognize the management and staff of the Mines and Geology Division (MGD) for the research support and technical assistance provided. Special mention is made of the MGD Summer Worker cohort of 2008 who conducted various aspects of the data collection and searches, and the author is extremely grateful. The MGD also permitted use and reproduction of maps and photographs in its archives and other repositories. Appreciation is also expressed to other entities and persons who have permitted usage of proprietary content as illustrations.

Special mention is made of colleagues Edward Robinson and Anthony Porter for their probing reviews, discussions, guidance, and motivation. There are many other persons who have also provided insight and direction for additional research, some of whom have moved on in this life, and even to the next. I recognize their invaluable support. I also express immense gratitude to the team at Springer Nature for their support and dedication in getting this work published.

Contents

About the Author

Suresh Bhalai has served over 20 years as a Geologist at the Jamaican Government's Mines and Geology Division. He started his career journey as an Engineering Geologist advising development planning and modelling natural disasters. In 2007 he was deployed as an Economic Geologist under the Government's mineral exploration program. In 2014 he was assigned to lead this program as the Director of Economic Minerals. Suresh has also temporarily served as the Principal Director for Mining and Minerals within the Jamaican Ministry of Transport and Mining both in 2018 and again in 2022. Here he provided technical advice to the political directorate, and leadership towards the implementation of policies for mineral management and use in Jamaica.

His training commenced at the University of the West Indies (Mona, Kingston) where, in 2002 he received a B.Sc. in Geology (First Class Honours). He later, under a Canadian Commonwealth Scholarship, pursued and received a M.Sc. in Mineral Exploration (2005) from Queens University, Kingston, Ontario. Suresh has balanced his technical training with programs for management and leadership. In 2015 and 2016 he pursued the Government's Public Sector Senior Leadership Development Programme; a high-level leadership training designed for senior executives tasked to drive Public Sector transformation.

Suresh finds fulfilment in sharing his knowledge, and he does this through his gift of teaching. He believes that one never is at a lost when we help someone to gain knowledge; because we would have made an impact, and it is then our lives become meaningful.

Abbreviations

AECA	American Economic Co-operation Administration
ASTM	American Society for Testing and Materials
BGS	British Geological Survey
BRGM	Bureau de Recherches Geologiques et Minieres (geological survey agency in France)
CD&W	Colonial Development and Welfare
CGS	Colonial Geological Surveys
CGSS	Colonial Geological Survey Service
CIDA	Canadian International Development Agency
DIPECHO	Disaster Preparedness for European Community Humanitarian Office
ERP	European Recovery Programme
FAO	Food and Agriculture Organization
FIDCO	Forestry Industry Development Company
GCE	General Certificate of Education
GSD	Geological Survey Division
IBM	International Business Machines
IDC	Industrial Development Corporation
JLP	Jamaica Labour Party
KSAC	Kingston and St. Andrew Corporation
MGD	Mines and Geology Division
MQD	Mines and Quarries Division
ODIPERC	Office of Disaster Preparedness and Emergency Relief Co-ordination
OPIC	Overseas Private Insurance Corporation
PNP	Peoples National Party
UNDP	United Nations Development Programme
USGS	United States Geological Survey
UWI	University of the West Indies

Insertions into quotations are indicated by square brackets '[ ]'
Abbreviated quotations, including extracts are marked by ellipses '. . .'
For numerical quantities the units used by the original source has been retained in most cases.

1 Introduction: Early Years of Mining and Geology

Abstract

All the settlers of Jamaica, beginning with the Taino people who came around 650 AD, to the Spaniards in 1494 and then the British in 1655 sought to capitalize on the country's earth resources. With the increase in demand and use of these resources to sustain lifestyles came the necessity for geological characterization. The development of mining and geological research is examined, along with the important role of Governments to provide objective technical guidance in the use of natural resources. Administrators oftentimes do not concur with the direction of scientists; however when they do, nations eventually advance.

1.1 Overview

The island of Jamaica became known to the Europeans just over 500 years ago with the landing of the Spanish. The Spaniards found a people, the aboriginal Tainos, who were already knowledgeable about the natural resources that the country afforded to support their modest lifestyle. The Tainos were the first to search and win the minerals they deemed useful. This required them developing knowledge of the earth resources and how to extract them. Further, they perfected techniques of beneficiation to process the material to a form of utility. With this said, the Taino people should be recognized and credited as the first to conduct geological research, mineral deposit evaluation and mining! Over time the Spaniards would continue this effort, followed by the British, and up until present, independent Jamaicans. Herein, we examine this history of the development of mining and geological research.

1.2 Land of Springs

The name *Jamaica* is derived from the aboriginal name *Xaymaca*. Spanish historians translated this aboriginal name from the Arawakan language of the Tainos which meant 'land of springs' (Black 1976) or 'an overflowing abundance of rivers' (Roxburgh and Ford 1901). The name reflects the Taino's perspective of the land, and this has survived the efforts by the Spanish to change it. This aboriginal group were the first to utilize the resources found in Jamaica; an effort that the Spaniards and the English would continue and advance in their exploration of the geological assets.

1.3 Life of the Taino

The Tainos are likely the first to inhabit Jamaica and are considered relatives of the Arawaks of South America originating from the Orinoco

S. Bhalai, *The History of Mining and Geological Surveys in Jamaica*, Historical Geography and Geosciences,
https://doi.org/10.1007/978-3-031-42604-9_1

Valley. They migrated from Venezuela and Guyana, passing along the Lesser Antilles and reaching Jamaica by 650 AD, where they established settlements in numerous areas. This people enjoyed a humble living and preferentially chose those areas endowed with basic resources necessary for their habitation. They chose locations having reliable fresh water supplies and suitable earth materials such as clay and sharp rocks for their daily use. They established themselves all along the northern and southern coasts from Priestman's River in the east to Negril in the west (Fig. 1.1). Villages were located as far inland as Ewarton and Moneague. The White Marl village on the Caymanas Estate near Central Village, St. Catherine was perched at the top of the hill commanding a view of the plains. At that time the Rio Cobre flowed at the base of the hill and entered the sea via the Fresh or Ferry River until a hurricane of August 28, 1722, aligned its course to the modern channel. The road to Spanish Town now traverses the site of the village (Black 1976; Jamaica National Heritage Trust 2011; Lee 2006).

Taino occupations included subsistence farming, fishing, hunting, trading, and artisanal pottery. This lifestyle was supported by tools and implements made entirely of natural materials. Technology was basic and implements for daily use were made of wood, shell, stone, minerals, or clay. Tools were made mainly from stone as iron was unknown. Stone tools such as axes and celts (implements with chisel tips) made from a green rock and to lesser extent a black variety have been found all over the island. Grey and blue schist, lava and some sedimentary rocks were occasionally used. These relatively hard stones were sourced from the Blue Mountains and at Green Bay, Port Henderson (Roobol and Lee 2006). The rivers such as the Rio Minho which drains the central interior of the island were also likely sources of materials, especially for the hard volcanic rocks which could be honed into sharp edges for celts. Chert, a locally common variety of silica with hardness like glass, was used as scrapers, knives, and awls.

There is some reservation toward the origin of some rock materials as the Tainos were traders. Similar rocks occur in Cuba and Hispaniola, and these Indians may have imported some of their tools. Christopher Columbus documented the skilful shaping of stones by these people, and he noted certain well-shaped, smooth, and beautiful stone implements placed in the bottom of water jars used to keep the water cool. Many of these have been found on the surface or unearthed in fields, and have been given the name *thunder balls* or *thunderbolts* (Black 1976). Hard rocks of local origin, such as porphyry, basalt, dolerite, and chert (Fig. 1.2), have been found near dwellings and kitchen middens (Zans 1955).

Beautifully designed and decorated clay vessels found common applications in daily living, such as cooking and serving, and storage of potable water (Fig. 1.2). The staple cassava was cooked on clay griddles or *buréns* (Rodney-Harrack 2006). Tainos buried their dead in caves, and sometimes they utilized a clay bowl for keeping the head and certain bones in secondary burials, and even food for the dead while on their journey to the afterlife. In 2019, while quarrying limestone in the Hellshire Hills of St. Catherine, miners found a shallow cave containing fragments of a ceremonial clay bowl, a four-inch long clay Zemi (religious figure said to house the spirit of the Taino) along with human bones from select body parts (Murphy 2019). Native bowl shapes of Jamaica were round or oval, with boat-shaped vessels being common. The Tainos also moulded clay into body stamps and designs used in ancestral worship (Rodney-Harrack 2006). Despite their heavy usage of earthenware, pottery wheel technology was unknown and the Tainos skilfully manufactured earthenware using a simple technique. Before firing, the clay was reinforced by mixing with sand, ash, vegetable fibres or crushed shell. The vessels were then shaped, and the walls thinned and compressed using a paddle-and-anvil technique. Clay was likely obtained from the river and stream courses along which they commonly settled. However, not all their earthenware was locally manufactured, and some may have been imported during trading.

The Tainos recognized the attractiveness of semi-precious minerals and would decorate

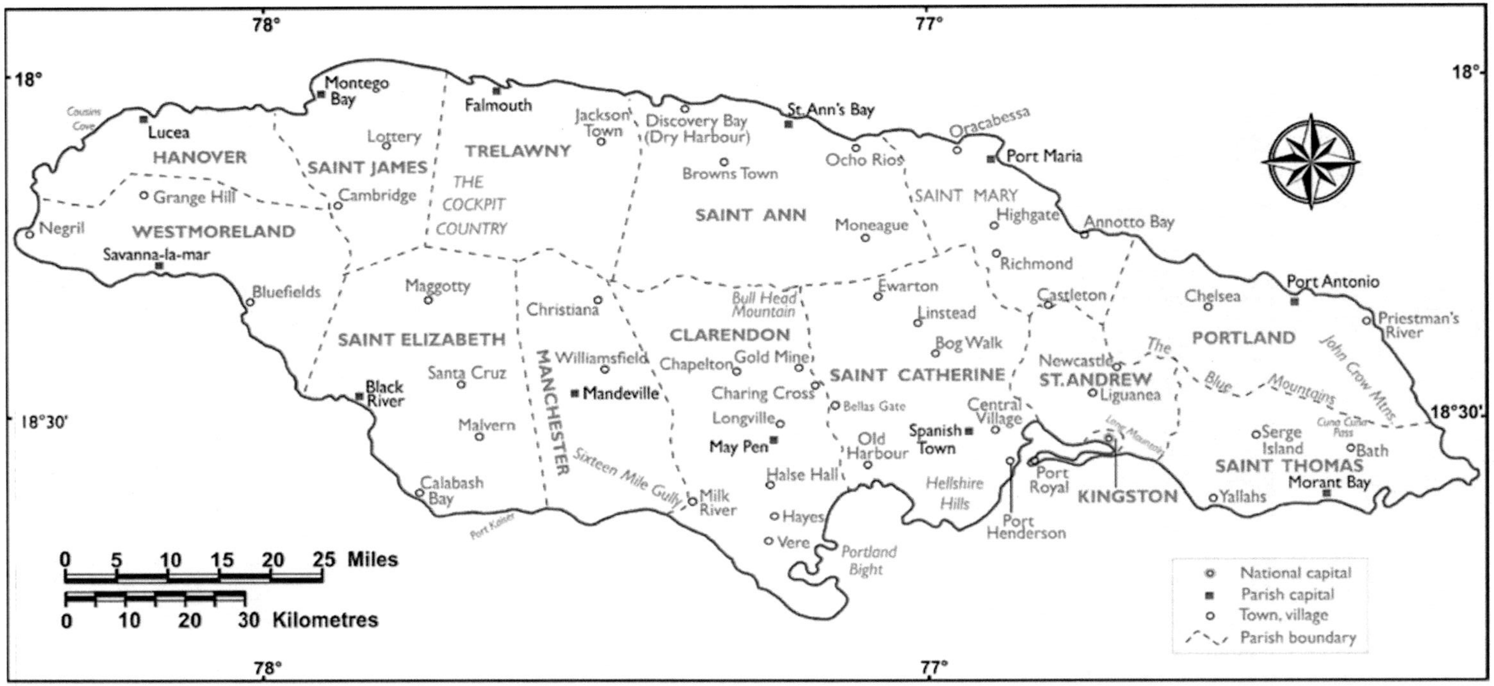

Fig. 1.1 Parishes of Jamaica (since 1866) and select places

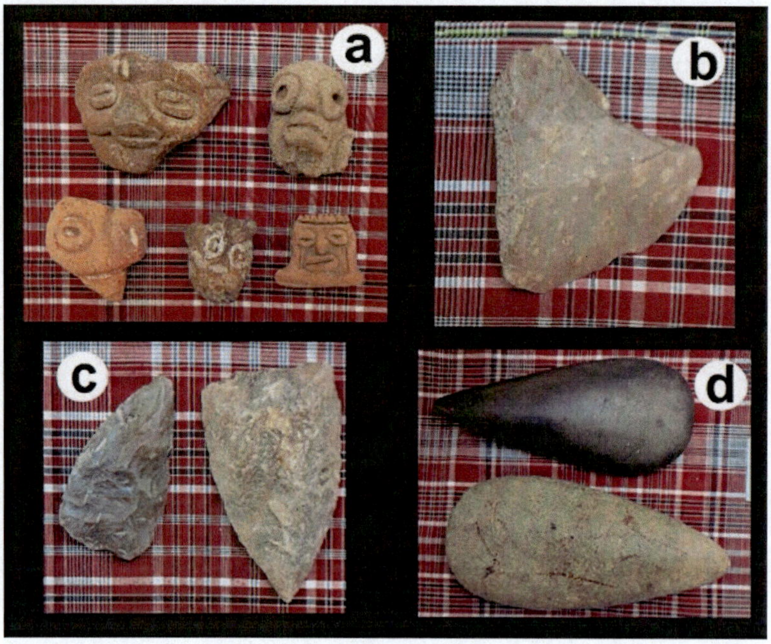

Fig. 1.2 Taino artefacts. **a**—Clay/earthenware decorations or 'Adornos' placed on vessels (larger sizes are 8 cm tall and smaller sizes are 5 cm tall) [650–1500 AD]; **b**—Chert knife (9 cm tall) used for cutting [650–1500 AD]; **c**—Chert arrowheads (8–10 cm long) which were attached to sticks and used for fishing and hunting. **d**—Stone Celts (20–23 cm long; the black one is made of basalt) used for chopping and digging, and in religious ceremonies [650–1500 AD] (Photographed by author at public cultural exhibition in Kingston, Jamaica, October 2022)

themselves with beautiful beads and pendants of skilfully rounded, polished, and perforated chalcedony and onyx. Off-white chalcedony beads have been found in St. Elizabeth and a bead of milky agate in Manchester. Other ornaments have been found in Vere, Clarendon. These are minerals originating from the central interior of the island and are found in certain streams and rivers, and on the coastline. Foreign stone pendants and beads made from coarse crystalline white metamorphic rock with black speckles (probably gneiss) have been found at Calabash Bay, St. Elizabeth. Ornaments made from this rock were also found in the southeast Caribbean. This rock may have originated from the South American continent. The Tainos also used white quartzite, not found in Jamaica, for sculpting images of their gods (Roobol and Lee 2006).

Stone ornaments were also worn with metal jewellery. At *Bahía de la Vaca* or 'Cow Bay', named by Columbus (now called Portland Bight), he met a Taino chief or cacique who was adorned in a band of green and white stone beads that supported a golden ornament. Two golden plates hung from his ears and a necklace of white beads had copper ornaments attached. The golden ornaments were made from a gold-copper alloy which the Indians called *guanin* (Black 1976). This, however, came from South America by way of the Eastern Caribbean, possibly by trade. The inhabitants of Peru and Mexico obtained gold from the sediments in riverbeds and those aboriginal groups of Hispaniola and to a lesser extent Jamaica may have also picked gold from the rivers by this method. In 1982, a small, 22.2 karat oval gold disc was found in St. Ann. It was a flattened gold nugget weighing 0.596 grams with some 92.5% gold, 7.2% silver, and some nickel and iron. The nickel content however, pointed to the alluvial gold-bearing gravels near the nickel deposits in the Dominican Republic. The ornament was donated to the Institute of Jamaica (Lee 1985).

1.4 Exploration and Mining by the Spaniards

On his first voyage in 1492, Columbus explored the Bahamas, north-eastern Cuba, and parts of the north coast of Hispaniola. The aboriginal people of Cuba spoke about an island in the south which abounds in gold. They gave the name of the land and indicated that it was the source of 'the blessed gold' (Gardner 1971; Black 1976). They may have referred to the trade relations they had with Jamaica where gold from South America was brought up along the Eastern Caribbean. Columbus came to Jamaica in 1494 and named the island Santiago (a name that failed to be retained). Despite numerous visits, he never found gold in abundance.

Columbus died in 1506 and his son Diego, through a legal battle for the titles his father had received from by the Spanish monarchy, was appointed Governor of the Indies in 1508. He came to Jamaica a year later and secured his interests and rights to the island by appointing Juan de Esquivel as governor (Black 1976). The King of Spain wrote to Diego in July 1511 that:

> Juan de Esquivel and the Christians in Jamaica should endeavour to find gold. If there is none, orders will be given as to how the colonists might live. (Cundall and Pietersz 1919)

A year later in 1512, the prospects of finding gold were deemed doubtful, and it was proposed to relocate the colonists to Cuba where they may have discovery success. The Spanish however, maintained Jamaica as an outpost for supplies, trading, and rest. Animal rearing, lumbering, sugar, and tobacco were more remunerative but some 'hatos' or ranches supplemented their income with small mines for metals, including lead, copper, silver and gold (Fig. 1.3). Some mineral discovery successes have been reported. In 1519 the Governor of Jamaica, Francisco Garay (tenure: 1514–1523) sent a piece of Jamaican gold to the King of Spain. This gold was likely won while processing local copper ores (Zans 1955). Copper mines were established in Clarendon and Portland, and lead was won from a mine at Liguanea (modern-day Kintyre, St. Andrew). Sawkins and Brown (1865) and Sawkins (1869) reported mines for gold and copper in eastern Jamaica in the Rio Grande and Swift River valleys in modern-day Portland parish. The Spaniards had mines for silver and copper at places now known as Epping Farm, Whitfield, Abbey Green, and a place known only as 'Spanish Adit', all in modern-day St. Andrew parish. The ore from these workings was transported by mules over the Blue Mountains into Portland where smelting was done at 'Spanish Mine' at a place called Leckies on the Western bank of the Swift River near modern-day Chelsea. Copper was also mined at Leckies. A mine is also reported as the mark of the eastern boundary of Hato de Morante near Morant Point, but there is no record of the metal won. This is likely copper extracted from the hills near Bath. It is recorded that some copper was found in the Healthshire (Hellshire) Hills, and this was used to cast the bells which hung in the old abbey church at St. Jago de la Vega (Spanish Town) (Gardner 1971; Siriunas 1993; Zans 1955). There is, however, no evidence of any mines in this region, nor any geological evidence that these hills have any economic metalliferous deposits.

Despite the economic activities, disease outbreaks had significantly crippled productivity in Jamaica driving the island into a depressed state. The Spanish King in July 1521 announced his awareness of the considerable loss and high death rate in the island caused by the general pestilence among the Indians (Tainos) and slaves (brought from Africa). He further indicated his willingness to aid development of the colony and was especially pleased that gold was discovered. The Colonists received an economic stimulus through a reduction of the taxes they paid to the Spanish Crown. The policy required that the retained funds should be used to recover from the losses incurred due to the loss of slave labour. The measure lasted for eight years and it included a concession on mining taxes where the tax rate was reduced from 20% of the gold to 10% (Edward 1942).

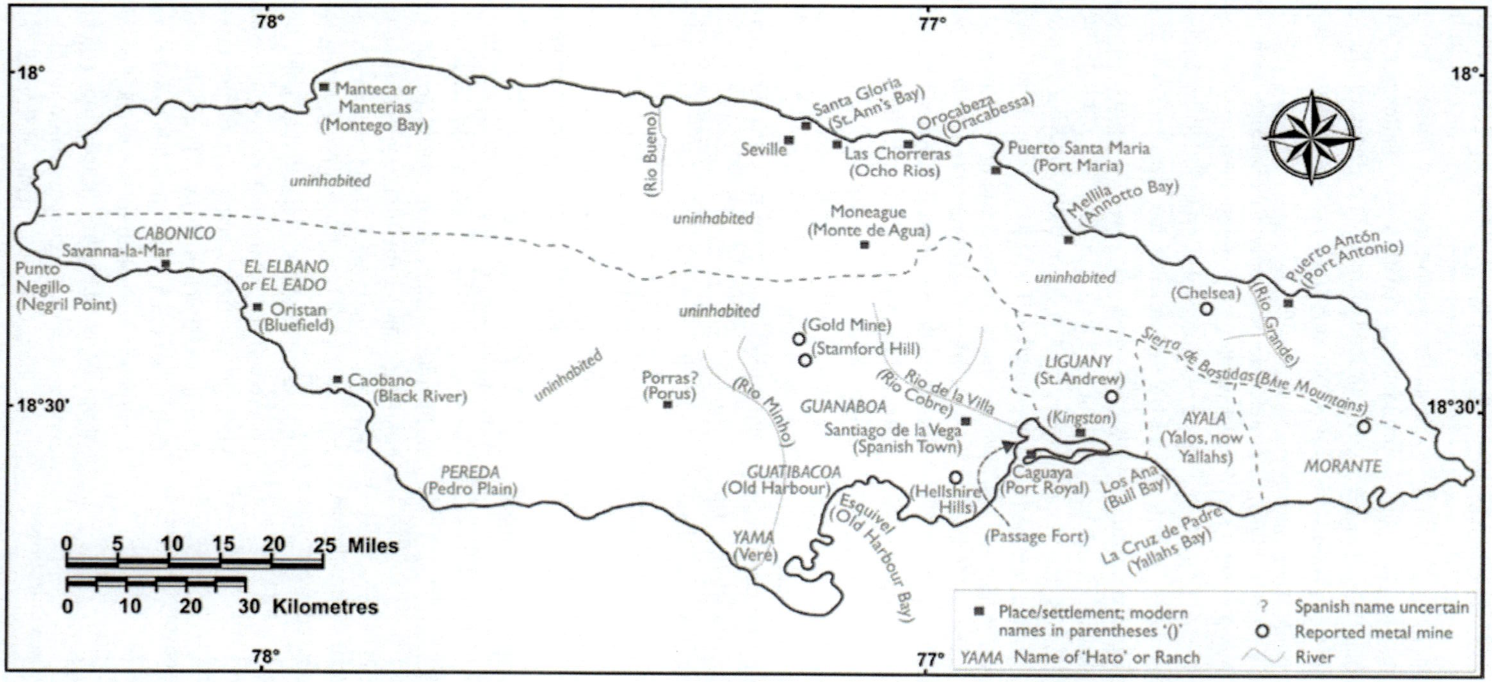

Fig. 1.3 Territories, settlements, and features in Jamaica during Spanish occupation (pre-1655). The Spanish monarchy had recommended to Columbus that the island should be called *Santiago* after St. James, the patron saint of Spain. However, the Taino name *Xaymaca* which evolved to Jamaica defied the efforts to change, and prevailed. *Guanaboa* and *Liguany* are Taino names. *Liguany* is now *Liguanea* and is now used to refer to the plains of modern-day Kingston and a small district in St. Andrew parish. From 1510 to 1535 the Island's capital was *Seville*. Except for a one-year period that started in 1755, *Spanish Town* was the capital from 1535 to 1872. *Kingston* (shown for reference) was the capital for that short period and since 1872. The more rugged northern side of the Island was largely uninhabited. 'Hatos' or ranches formed the territorial divisions and were established on the plains and less rugged landscape on the south side. Cattle, horses, and pigs were reared and there were sugar works, lumbering forests and mines for gold, silver, copper, and lead. The old lead mine at *Liguany* (at modern-day Kintyre) and the gold mines at Gold Mine and Stamford Hill were worked by the British. Gold was won on the north of the Sierra de Bastidas (Blue Mountains) and transported on mules over the mountain through the Hato of Morante and shipped from the small port at Caguaya (Port Royal) [Compiled from information presented in Black (1976), Bryan (1992), Gardner (1971), Padrón (2003), Prestwidge (2008), Sawkins and Brown (1865), Sawkins (1869) and Zans (1955)]

The records in the Seville archives documented productive mining ventures in Jamaica and the associated fraud. One manuscript of July 1519 includes instructions to a Jhoan Lopez de Terralba from the King that upon appointment to the post of Comptroller of the Island of Santiago called Jamaica, he should audit the account of Pedro de Mazuela, the Treasurer, and hold him responsible for the collection of dues payable to the Spanish Crown by smelting houses, ranches (hatos) and farms and other avenues of profit and custom duties (Edward 1942). His report should include the amount of gold produced. A decree issued at Toledo in December 1525 noted the appointment of Luis Sanchez de la Torre as the supervisor of gold smelting and marking in the Island of Santiago called Jamaica. On the recommendation of Luis Columbus, son of Diego, the King, declared that all gold and silver extracted in Jamaica should be marked with the royal arms of the Crown of Castilla—León (castles and lions) and around the escutcheon the word 'Jamaica' (Black 1976; Edward 1942).

The metalliferous potential of Jamaica was insignificant when compared to the rich mines in Mexico and the South American states. However, some mineral deposits, though small, showed potential to be profitable and were encouraging for further exploration work. The old, exhausted mines that were once productive were especially attractive. Abbot Francisco Márquez de Villalobos reported in 1582 that he lost five hundred Negro slaves imported from West Africa in place of the Indians, in opening *'las minas viejas de oro que lo hay muy bueno y rico'* (Zans 1955). This abbot, who came a year before, was the first to live in Jamaica, and he had written then that, 'Previous abbots had had more concern in making incomes than in attending to their duties' (Black 1976). Governor Fernando Melgarejo de Córdova is noted for ordering exploration for copper in an area known only as the 'Valley of Upar'. In a letter to Spain dated May 1598, he wrote about copper he was sending to Cartagena (Colombia) for the founding of artillery. He further wrote in 1601 about the 'Bastidas mountains' (the Blue Mountains), some 28 leagues from St. Jago de la Vega (Spanish Town), where some rebel Indian lived and there is much gold. He indicated that, at his own expense, he has dispatched an exploration party headed by Francisco de Castro to find the rebels and to discover gold mines. Intuitively, the rebels would be useful for providing directions to the gold deposits (Gardner 1971; Padrón 2003; Edward 1942; Zans 1955).

By the 1600s the Jamaican mines were reported to be exhausted. Jacinto Sedeño de Albornoz who had come to audit the country's accounts reported to Spain in 1655 that there were many gold mines in Jamaica evidenced by the old pits. He further noted that much gold of high quality was taken. He however linked the collapse of these mining operations to the terrible treatment of the Indians even though labour from West Africa was imported to offset the loss (Cundall and Pietersz 1919). However, the mines in Jamaica were mostly small-sized, and the miners were likely extracting the best ores first (or 'high-grading'). This would initially give greater remuneration but at the expense that mostly low quality ore remains, which would be uneconomic to process, and the mine unprofitable.

Today, there are some place names in Jamaica derived from the Spanish language which have survived. Many were assigned to be descriptive of the surrounding environment, and some included the words for gold and copper such as *Rio d'Oro* and *Rio Cobre*, two rivers in St. Catherine parish. The latter flows through Spanish Town (once referred to as Villa de la Vega) and was then called *Río de la Villa*. This river was named by the British who saw the word copper pointing at a location near the river and thought that was the name of the feature. Oracabessa was once called *Cabeza de Oro,* but this referred to a nearby hill which was covered in yellow flowers at a certain time of year. The *Rio Minho* in Clarendon parish on early maps (Bellin 1758) is stated as *Rio Mino* and may be a corruption of 'mina'. The river drains the mountainous hinterland where gold was found and mined. On the bank of the Rio Minho near the Longville Estate, at May Pen a Spanish gold washing apparatus termed *lavaderos* was also identified (Long 1774a). It was observed that when the river flooded, particles of gold could be found in the sediments and the Spaniards adopted this crude gravity separation process from the Indians of Peru and Mexico. The device consisted of a terraced platform with several connected basins hollowed out of rock and rendered with mortar to obtain a smooth surface. A large gutter was connected to this apparatus to transmit river water to the basins to wash the sediments. Upon locating a paystreak of gold, the stream would be diverted, and the rich sediments collected and subjected to repeated washings until the grains were separated. The Indians standing over the apparatus would scrub the material of rocks and plant matter. The concentrate had gold, some sediments, and black sand. Further washing would separate the valuable commodity (Long 1774a; Gardner 1971). In the early years of English occupation of Jamaica, the estate manager of the Longville estate brought the owner, Alderman Beckford, a very tiny piece of gold he found in the sediments shortly after the Rio Minho had flooded (Long 1774a). The manager encouraged his boss to deploy an exploration party to search the hills for the source of the gold. However, Beckford was unmoved. He simply pointed to the cane fields and explained that with such a mine of wealth on the surface, it would be a waste to search for one beneath it (Gardner 1971).

1.5 The English Conquest

A civil war had resulted in England becoming Commonwealth in 1653 and Oliver Cromwell was declared the Lord Protector of England, Scotland, and Ireland. The following year he ordered Admiral William Penn and General Robert Venables to initiate attacks on large Spanish territories in the west. A failed attempt on Hispaniola forced the English army to attack Jamaica, an island known to be sparsely populated and weakly defended. By this time however, Jamaica had lost importance to the Spanish. The island had mostly ranches, and the better gold and copper mines were already exhausted. There were about 1500 colonists and only a third were trained to fight. On May 10, 1655, 38 ships with 8000 men landed at Passage Fort in Kingston Harbour. The Spaniards were given surrender terms which allowed them time to leave the island. They cleverly used the opportunity to free their animals and move their belonging to the mostly uninhabited north coast to depart for Cuba. When the English entered Spanish Town to their disappointment it was deserted with no spoil. In anger the churches were burnt and the copper bells in the old abbey church were melted for shots (Black 1976). These bells, as Gardner (1971) reported, were cast of copper which was supposedly mined from the 'Healthshire' (now Hellshire) Hills.

Cromwell, though disappointed with the defeat at Hispaniola, was willing to develop Jamaica. He issued his '*Proclamation giving Encouragement to such as shall transplant themselves to Jamaica*'. This addressed the granting of land to settlers and the rights to be enjoyed by persons engaged in fishing, mining, quarrying, among other activities. There was also a three-year custom duty relief. The Proclamation also declared that all persons born in the island shall enjoy the same rights as citizens of England. It allowed the colonists the ownership of all minerals, except for gold and silver, for which extraction probably would have to be negotiated:

> That they shall hold and enjoy all, and singular mines of copper, iron, tin, or other minerals whatsoever (excepting gold, and silver mines) and all mines of quarries, coal, stone, alum, or other mines, whatsoever (except as aforesaid) within the circuit, meets or bounds of the said several and respective proportions of land; . . . and also full power, and authority to man, and send forth to sea, and unto any the coasts, and shores, roads, harbours or creeks within or near the said Island, any ships, boats or other vessels to fish for, find out, or take any pearls, precious stones, or jewels therein being, and to enjoy the same to his and their own use or uses, rendering and paying to the Governor of the said Island for the time being, or to such other person or persons, for the time being, as His Highness shall authorize to receive the same, to His Highness use, the full fifth part only, and no more of all such pearls, precious stones and jewels as shall be got, found,

> and taken, as aforesaid; and also one tenth part of all such metal as shall be had, found, and gained in the mines, granted hereby to the aforesaid Planters.
>
> —Oliver Cromwell, Whitehall, October 10, 1655. (extract from Cromwell 1655; Sec. 1.6)

Settlers in New England were also encouraged to relocate to Jamaica (Gookin 1656). There may have been some initial reluctance to relocate and as such a thousand Irish girls and some Scotsmen of questionable character were brought to strengthen the Jamaican population (Black 1976). The European settlers were not readily adaptive to the tropical climate and so, upon recommendation to Cromwell by Commissioner, Major General William Brayne, the importation of African slaves began. Jamaica was maintained under military rule for the first six year of English occupation (Black 1976).

In 1658 Cromwell died, and by 1660 the monarchy was restored with Charles II, the son of Charles I who was defeated in the civil war. The colonists of Jamaica were uncertain if they would still enjoy the rights bestowed by Cromwell. The new king, however, expressed that development of the island would continue as started by Cromwell. Cromwell had placed a colonel from the army of invasion Edward D'Oyley in command of Jamaica, but Charles II replaced him with a young new governor, Lord Thomas Windsor. Windsor brought with him the island's seal, a mace and a Proclamation by the king dated December 14, 1661 (Black 1976; Charles II 1661). This Proclamation promised the English subjects in Jamaica the same rights as those enjoyed in England and further increased the portion of lands granted to each settler. It also reduced the royalty on metals by fifty per cent. Gold and silver remained the possession of the Crown:

> . . . and all copper, lead, tin, iron, coals, and all other mines (except gold and silver) within such respective allotments, shall be enjoyed by the grantees thereof, reserving only a twentieth part of the product of the said mines to Our use – Charles II, Whitehall, December 14, 1661. (Charles II 1661)

Mining interests had to apply to (and probably negotiate with) the Crown to mine precious metals. An example of this is a mining proposal recorded in the Minutes of the Legislative Council of Jamaica where a petition is presented to the Council in 1744 from merchants William Perrin (who owned lands in St. Andrew) and Matthias Philip to conduct mining for gold and silver (Edward 1942). The men requested of the Crown special mining concessions and reductions in the royalty reserved by the Crown on any mines found. The merchants proposed a royalty of 6% of the gross ore or 5% of the gold and silver that can be extracted. Their request was approved, but the success of their venture is not known.

1.5.1 Discovery Failures

The English recognized that many of the old Spanish mines were exhausted and that the richer veins were already worked. Sir Hans Sloane, in 1707, remarked that the Spaniards knew of gold and copper in the region called Liguanea (Edward 1942). Liguanea, before 1655, encompassed present-day Kingston, St. Andrew, and a small part of Portland (Anonymous 2011). Sloane recorded that a Sir Thomas Lynch questioned the former Spanish inhabitants of Jamaica who had fled to Cuba, about places where they had established mines and they told him of Liguanea. They, however, stated that they had not worked them (Edward 1942). Despite the seemingly positive leads from the old settlers Lynch never found these deposits.

The Spaniards had worked several productive mines for mostly copper and some silver in the hilly interior of Liguanea, and the ones they mentioned may have been the small and unimportant deposits. The Spaniards had worked the Hope lead and zinc mine at the foothills at present-day Kintyre from about 1580. This deposit had lead, zinc, silver, gold, and copper, and it is likely that the lead produced may have been used for shot during the wars in the Caribbean between the English and Spanish (Siriunas 1993). Browne (1789) documented the various metals associated with the Hope Mine and he described the lead ore as being rich, but the very fine-grained nature of the rock demanded longer milling and processing time. The silver, he stated, occurred in irregular veins. He also gave an impressive list of fourteen copper minerals. Abundant sulphur in the ore presented metallurgical challenges such as contamination of the leaching reagent. Antimony associated with the copper minerals also complicated extraction. He explained that these factors contributed to unprofitability of the mine which led to the venture being disposed of by the operators. The proprietors had beforehand incurred great expense in outfitting the mine with complete works for ore processing. Both the accounts of Browne (1789) and Long (1774a) suggest that there was continued development of the Hope Mine in the 1700s.

British colonial administrator and historian, Edward Long, wrote that '*the mountains abound with copper ore…, lead likewise abound here*' (Long 1774a). He reported that two English men were able to open and work two copper mines in the Liguanea hills, but these became unprofitable. One mine is said to have produced no less than a tonne of metal per week with the failure attributed to low capital investment or lack of expertise of the proprietors (Long 1774a). Browne (1789) mentioned two mines: one owned by Sir Simon Clarke and the other owned by a Mr. Anderson. Clark's mine, located in St. John, was rich in the blue copper carbonate, azurite. He had spent a considerable sum, but the attempts were ultimately abandoned (Gardner 1971). Anderson, he stated was not experienced in mining and often extracted with great difficulty fine-grained pyrite and mistakenly discarded the good ore. He concluded that lack of experience in proper identification of the metalliferous minerals was a common problem among the miners.

Despite gold being more profitable, some prospectors preferred to search for silver instead. Browne (1789) recorded that two or three Jewish prospectors had obtained a mining patent to locate an old Spanish silver mine in the 'Healthshire' or Hellshire Hills, St. Catherine with the intention to operate. They based their project on a story that the Spaniards had found silver there. This mine was never found (Gardner 1971).

Many of the mines in Jamaica failed and Long (1774a) described that the concessionaires gladly parted with their operations and surrendered their mining patents. Browne (1789) optimistically considered that the colonists should continue their search as the demand existed for a supply of metals. Estate necessities such as ploughs, mill equipment, copper pots for the boiling house, stills, carts, hoes, and cutlasses were all produced from metals. A planter he stated oftentimes would have to acquire this equipment by credit from English merchants. Therefore, assisted by local planters he undertook detailed analyses to demonstrate the profitability of local copper and lead mining to supply the plantations. He computed that a sugar estate producing one hundred hog's heads (the large barrels in which sugar was shipped having filled weight of about 813 kg) per year would spend £65 each year on copper and lead. This he translated to £23,700 per year; money that could be retained in the colony if the metal was produced locally. He also stated that the cost of the metals would be much lower, and the mines would be a source of employment, resulting in the strengthening of the colony.

English historians such as Long (1774a), Browne (1789), and Renny (1807) ascribed greater mineral discover successes to the Spaniards who had experience working similar deposits for over 2000 years in Spain and Portugal. In Jamaica, the Spaniards worked the richer deposits exploiting the better grades of ore. This posed a challenge to the English who it seems were inexperienced in mining and inherited the poorer quality ore. Like the Spaniards, the English also had a major challenge with fraud. Some English miners entered ventures that they knew were unprofitable and would only achieve success with the intervention of large mining companies but were still pursued to exploit the capital invested (Long 1774a). Contrary to Browne (1789), Renny (1807) identified that it was more profitable for the landowners to engage in agriculture than investing in mines of which they had no experience.

1.5.2 Other Mineral Resources

Materials such as stone and clay were also explored, and greater use of the local materials was encouraged. Some locally available resources were utilized but others were substituted with imports. White limestone found all over the island was commonly used for dimension stone in construction projects, or polished as marble for use in buildings. Soft, fine-grained free-stone was obtained from a large quarry near St. Ann's Bay and used in the elaborate plantation great houses such as Seville. It was also used for carving decorative elements affixed to columns and walls to enhance their beauty. A similar variety was found in St. Thomas-in-the-East (Browne 1789). Black and white varieties of natural marble found in the hilly interior of eastern Jamaica were also identified, but up to the late 1700s were not quarried. The deposits were in remote areas and a competitive product from Bristol, the primary supplier for stone was more economical. Stone from Bristol was used in the churches and large buildings (Long 1774b). Browne mentioned that a smooth even-grained whitish 'bastard' marble from the mountains east of Kingston (probably a low-grade marble in the Blue Mountains) is confused and sometimes used as a substitute for limestone. In addition, a very cheap lightweight sandy free-stone was also imported from Bermuda for use in Jamaican buildings (Browne 1789).

Material for road construction was sought from the riverbeds and shingle deposits on the coastline. Hard but smooth rocks were used as cobbles to improve the ride and provide firm support to heavily laden wagons that would become stuck in muddy roads when it rained. The shingle beaches of the Palisadoes was a convenient source to supply the heavily trafficked routes. However, in 1716, an Act of Parliament was passed forbidding the removal of shingle from certain parts (Steers 1940) as the removal promoted encroachment of the sea which eventually would damage the road to Port Royal.

Jamaican white limestone was also found to make quality lime and cement used in construction. Small-scale lime burning was done across the island but in non-commercial quantities. Local lime was found to be of equal quality to that manufactured in Bristol, the lead supplier for Jamaica. The sugar estates used this chemical for building purposes and purifying the sugar, and it was observed that using locally produced lime resulted in higher-quality refined sugar. The planters, however, preferred to import the more expensive lime from Bristol, rather than bother to burn their own. It was suggested that the poorer class of settlers should set up lime kilns and establish a fixed market for their products by supplying the sugar estates (Long 1774b).

Clays, phosphates, and salt were also important. Local clays, such as from Spanish Town, were utilized for earthenware, water jars and vessels, and tiles. Seams of white clay found with the reddish clay at Sixteen-mile-walk (now called Bog Walk) were suitable for making tobacco pipes (Browne 1789). The limestone caves of St. Ann hosted nitrate-rich bat guano with quoted reserves of 'many dozens of cart-loads' that could be used for fertilizer (Long 1774b). Sea salt was a major commodity of export from the island in the early days of colonization. Browne observed that all the salinas or salt flats were lightly draped with salt, and he noted the good potential for large quantities of high-purity salt to be manufactured. High labour cost however was the limiting factor (Browne 1789). The importance of the commodity declined towards the end of the 1700s when it seemed that labour costs were increasing. Long (1774b) described the salt works on the four-foot deep Great Salt Pond on the southwestern end of the Port Henderson Hills along with two others in St. David parish, operated by Captain Joseph Noye. Captain Noye had produced in one year, ten thousand bushels of salt from these three operations but reported that he could have made many tonnes if there was an opening in the market for the commodity. Long (1774b) noted that the large-scale salt production at Tortuga suppressed prices and forced closure of many of the Jamaican operations.

An unusual practice of using mineralized water to make hones (whetstones) is documented. The waters draining from the limestone regions and having petrification properties were said to be used for making a hardened block suitable for sharpening razors and knives. Hardwood was carved to shape and then soaked for months in the special waters flowing from certain mountain springs. The dissolved minerals would accumulate as stony particles deposited between the wood fibres. This theory is not plausible, as petrification takes a much longer time and calcareous material deposited from the water would be too soft for sharpening iron tools. Grinding stones, particularly those used on the plantation where iron tools were common, were supplied by Bristol (Long 1774b). Another odd story was the use of a calcareous soil called *aboo* earth or clammy marl. This smooth, greasy, and cohesive soil found as seams in marl beds was eaten by the slaves as a pastime where they enjoyed a sweetish taste. Browne (1789) recognized its toxicity when used consistently over long periods where it entered the bloodstream and obstructed the minute capillaries. The clay accumulated in glands and smaller vessels of the lungs. He concluded that the slaves ate it to satisfy sharp stomach cravings either from hunger, worms, or some unnatural habit of the body.

1.5.3 Copper Boom

In the mid-1830s the copper deposits in Jamaica were attracting the attention and investment of the capitalists in Britain. The excitement associated with this possibility for wealth corresponded with the downfall of Jamaica's plantocracy and the corresponding decline of the the country's agrarian economy. This energy focus lasted only a few years and waned by the early 1840s (Stafford 2002). However, the momentum generated by the 1849 California Gold Rush generated renewed interest where mining was being considered as an economic alternative to agriculture, and mineral discovery successes would attract much-needed investment to the island. Gold finds were announced, such as that discovery in the early 1850s by an ex-slave who found some gold on land he had purchased on the Peace River coffee plantation in Clarendon (Anonymous 1852). Copper, gold, and silver exploration resurged and mine development in Jamaica experienced a short-lived boom. Copper deposits were more common, and these were primarily explored. Isaiah Deck, mining geologist from New York came to the island in 1854 and spent several months evaluating the mineral deposits. His findings were published in the popular *Mining Magazine* (Tenney 1854), a leading mining industry publication with extensive readership. His visit and the alluring report prepared from it were likely funded by local miners and owners of mineral-bearing lands.

Deck spoke quite convincingly of the copper mining districts in the central and eastern parts of Jamaica. He presented the island as a mining haven where small enterprises in the extractive industry could be sustained profitably. He noted that all the mining properties examined encompassed four thousand acres of rich agricultural and plantation land in the parishes of Metcalfe, St. Andrew, and St. Thomas-in-the-Vale; all within reasonable distance from the main shipping ports of Kingston, Port Maria or Annotto Bay. Other services available to the mining districts included the transportation network that was manoeuvrable by carts and wagons directly to the respective wharf from which shipments could be made to any American port. He identified that waterpower at the mines was continuous and abundant to sustain the heavy-duty crushing mills and concentration plants. The mines also required much timber for the buildings, works, and shaft support, and these were quoted as plentiful and of good quality.

He spoke favourably on the recommended methods for developing a mine, noting that the ore bodies occurred on level areas or on the sides of hills and gullies. Easy access he said could be gained by adits, side levels, and galleries without having to sink shafts which would require a headframe and lifting equipment. During the wet season, drainage from the surface would pool in the mines as the rocks were highly fractured. This was a problem, but Deck optimistically advised that the adits could be aligned to drain them sufficiently without using pumps. Deck relied on the geological work of Henry Thomas De la Beche to guide him on the various rock types. He declared that Jamaica had all the '*metalliferous strata*' typical of the best ore fields in the world and sought after by the companies. The rocks he stated are like those in Cornwall, England, and in Australia where experience has proven them to have productive mines.

Deck had examined with detail, twenty-five mine openings that would be developed or were in production. He chose fourteen as being promising, easily worked and at the state of completing the payback for the development capital expended. These included Dublin Castle spanning five hundred acres in Metcalfe, Mount Royal, Providence, Belmont, Chesterfield, and Mount Airy. He also examined Mount Ogle, which was the closest to Kingston, where development work had just started and seemed a 'most favourable indication'. The Dublin Castle deposit he stated had four lodes in a good location for working as there is an abundance of waterpower. He projected that one of these mines will get richer with depth where already 15 tonnes of excellent quality ore had been extracted. At Mount Royal where there were three lodes, there was 'an excellent vein' and another 'rich vein' having material 'yielding at least 50 per cent of copper'. He thought the Mount Royal veins to be the best and emphasized preference of these over the others. He indicated that they would immediately pay, yielding between 40% and 50% of copper. Deck noted that many of the mines were sufficiently developed to start ore winning and making profitable returns. He specified that some will indeed pay liberally, while others are at the point of just clearing their expenses. For labour, the mines faced a reasonable outlay of US $0.50 per day without boarding or lodging. He stated that this was better than in the USA and that the native labourers were quite satisfied, thinking themselves to be highly paid. Deck added that at this rate the mines could employ as many persons as they need and that a few white 'overlookers' (foremen) could be hired to keep things in order and deter the natives from certain undesirable working habits.

In summary Deck estimated that all the fourteen deposits he considered favourable could produce between 36 and 40 tonnes per month of dressed ore having grade of 36% copper. This would be worth over US $200 per tonne, not considering the occasional high-grade nodules of ore having over 60% that are found in some lodes. He considered it a rule-of-thumb that as the veins extended in depth, their tenor increased. The labour for this rate of mining was estimated at thirty to forty men including the foremen. With this expense the mines would remain largely profitable as the ore can be moved directly from the mine to the copper trading centres of New York or Boston, at a cost of transport and freightage of under US $20 per tonne. The first shipment of ore from these veins was made to the Revere Copper Company of Boston, some months before his visit to Jamaica. When Revere smelted the ore, they reported that it was 'of very desirable quality and yielded 24.5% of copper'. This ore was mined using local labourers and shipped undressed and mixed with gangue (waste) material. Deck compared the English ores of Cornwall, Devonshire, and Staffordshire to be of lower quality where the grades rarely exceeded 8% or 10%.

Deck optimistically stated that the deposits he assessed showed:

> . . . positive evidence of their being [*or, that they are*] permanent veins, or in the immediate locality of some immense main source of copper, as indications are numerous over the range of country. . . surveyed, and they have been traced over many thousand acres during examination. (Tenney 1854)

Deck's encouraging report resulted in increased enthusiasm and further investment in the potential copper paradise that existed in Jamaica. Prospecting and mining boomed and in some places were improperly planned and hit-or-miss. Shafts and adits were dug in hundreds of different parts of the island including the known mineral districts of the Blue Mountains and Port Royal Mountains to the east, and the Bullhead Mountains of upper Clarendon in the central parts. While some openings were made in metalliferous districts, there were mines sited in places known to be barren. It was an opportunity for the honest and the deceitful. Various companies were fortunate to ship several hundreds of tonnes of copper ore, but ignorance of local geology coupled with inexperience in mining led to their demise (Scotland 1889).

The Henderson's Jamaica Almanack of 1855 reproduced in Sawkins (1869) listed several mining ventures. Among these were some stable operations that had started before this mining boom. Registered in London were Clarendon Consolidated Mining Company, Port Royal, and St. Andrew's Copper Mining Company, Metcalfe Mining Company, and General Mining Company. Clarendon Consolidated mined silver, copper, and gold from Stamford Hill and Gold Mine in Clarendon. Up to the 1860s this company was able to ship dressed ore to England (Siriunas 1993). The Mount Vernon Copper Company commenced mining in 1836 during the initial short-lived copper boom but suspended operations then resumed in 1850 extracting copper from Lucky Valley, eastern St. Andrew. The Wheal Jamaica Copper Company started in 1854 with £100,000 capital and operated the Charing Cross Mine, Clarendon (Zans 1955). Over 207 tonnes of copper concentrate was shipped by the company. The first shipment of 4 tonnes with up to 12.5% copper was sent to New York in December 1854, probably for smelting tests. After that, from October 1855 to December 1857, the company shipped 203 tonnes to Liverpool having an average of 14% copper but sometimes up to 24% copper. According to the Decordova Price Current of December 31, 1857 (quoted in Sawkins 1869), this ore had a value of £2,817 of which the New York shipment was valued at £43. Some discarded specimens of ore from Charing Cross left at Halse Hall Great House were found to be rich in gold, with native gold being visible. Gold ore at that time fetched between £60 and £70 for each tonne not considering copper grades of between 12% and 15%. The price of copper concentrate varied and at times fetched as low as £9. The highest value fetched per tonne however was £25 at 21% copper grade. (Sawkins 1869). The Charing Cross copper deposit had copper mineral veins up to sixteen inches wide. It is stated that if the mine was operated by an individual rather than a company, greater attention to detail would have been given, and the gold potential may have been noticed considering that native gold was commonly mistaken for chalcopyrite or pyrite (fool's gold) due to poor skills in mineral identification (Sawkins 1869).

Ore was also taken from the works at Stamford Hill, Friendship, and Gold Mine in Clarendon. The Charing Cross and Stamford Hill mines intersected water at their lowest levels, and it became necessary to drain the openings. The options were to either invest more capital to pump out the water, or simply to abandon the works. The latter was chosen, and Scotland (1889) reported that 100 tonnes or more of dressed ore worth at least £1,000 were left in one of these mines. As the water level increased the mines became totally inundated. In modern times water continues to drain from the adits at Stamford Hill.

Copper was also mined in Portland by the Rio Grande Mining Company at Providence and Rose Hill in the east, and at Coopers Hill in the Rio Grande Valley. This was shipped to the USA. Lewis Hoffman, Government chemist of Jamaica, analysed samples from Portland and found native copper, cuprite, chalcocite, bornite, and chalcopyrite. The average grade of the deposits was about 62% copper; sufficiently high for the US smelters which were incapable of handling ores of lower tenor. At Phillips Gully and Bloxburgh in St. Andrew in 1860, the Jamaica Copper Mining Company, run by Alexander Fyfe, sought to extract primarily cobalt, and copper as by-product but this mine was unprofitable (Hose 1950 and Siriunas 1993).

The Hope lead–zinc mine at Liguanea was reopened in 1856 by the owner of the Hope Estate, the Duke of Buckingham under the name Hope Silver–lead and Copper Company. Soon it was the leading mining venture in the whole island where ore was shipped from Kingston to England. The mine was worked at six levels with nearly 5600 feet of shafts and adits. Grades of nearly 2% lead, 4% zinc, 0.2 troy ounce per ton silver, and 0.1 troy ounce per ton gold were obtained from this mine. Fifteen thousand tons of ore are said to have been extracted, being won during development excavations, as modern investigations have shown that very few of the actual stopes were mined (Siriunas 1993). The ore was crushed and concentrated before being shipped. The concentration plant is inscribed with the year 1858, a record that the facility was upgraded, with the optimism of prolonged production. Mining ceased in 1860 possibly due to metallurgical problems as it was long known that the ore contained much antimony which proved challenging when concentrating the material (Browne 1789). Districts such as Tomsfield, south of Annotto Bay, St. Mary were also mined for lead. Unlike Hope Mine this was not quite productive, and the operation was abandoned in 1860 (Zans 1955).

The local mining boom lasted from 1850 to 1858 with such positive outcome that public interests in mining flourished. Scotland (1889) described the vigour of this period where:

> Almost every land-holder in a promising district, down to the three and five-acre settler had signed away his underground rights to some mining company, or agent or enterprising speculator. (Scotland 1889; 17)

Using the illustration of the rich copper showings at Jobs Hill, St. Mary, Scotland (1889) further commented on the abrupt boom and bust of the mining projects:

> Job's Hill mine in Metcalfe, came very early to the front, first to fan the flame of excitement, and then to smother our mining enthusiasm by a collapse. A landslip after a flood revealed apparently a whole mountain of rich copper ore. Job's Hill shares therefore went up to a fabulous premium. Three or four hundred thousand tons of first-class grey copper ore were easily extracted. But the mining experts were from the first, suspicious of Job's Hill (Scotland 1889; 18).

The copper boom peaked around the mid- to late 1850s then declined sharply. Some investors, however, were willing to pursue and so they hired an experienced Cornish mining captain, Jehu Hitchings to report on their mines. The Cornish were known for their mining expertise especially in working tin and coal deposits. Hitchings visited around 1857 and Scotland (1889) illustrated the findings by stating that mining ventures '*came down by the run, like a collapsed parachute*' (Scotland 1889, 20). The year 1858 marked the bust of metal mining, and by 1862 the momentum had waned (Zans 1955).

1.6 Early Geological Research

It was Englishman Henry Thomas De la Beche (Fig. 1.4a) who, by persistence, demonstrated to the British Government the relevance of geological surveys to economic development. Though not the first Government geologist, he conducted systematic mapping of England simultaneously with the Ordnance Surveys (Wilson 1985). De la Beche visited Jamaica in his youthful years and conducted a geological survey mainly as personal research (Sharpe 1997). This was a first in the Western hemisphere. He assembled his findings to produce the first geological map of the eastern side of the Island to correlate the Jamaican rocks to those in Europe on the assumption that all the land masses were once connected (Sharpe 1997).

Before De la Beche, dedicated research on the geological make-up of Jamaica was lacking. Insight to the geology, more so the physical landscape and the use of minerals were commonly documented in personal accounts of Long (1774a, b) and Browne (1789). Of general scientific curiosity was Barham's (1718) letter describing a meteorite impact in Jamaica.

1.7 Henry Thomas de la Beche

Henry Thomas was born on February 10, 1796, in London and descended from the line of Thomas Beach I, a descendant of the Lords De la Beche of Aldworth in Berkshire who was the Attorney General of Jamaica and later the Chief Justice. Thomas Beach had married Helen

Fig. 1.4 **a**—Sir Henry Thomas De la Beche, geologist and owner of the Halse Hall Estate, Clarendon, Jamaica, and the first Director of the Geological Survey of Britain; **b**—His Excellency, Sir Kenneth Blackburne, Governor of Jamaica unveiling a plaque to recognize the geological work of Henry De la Beche at the Halse Hall Great House, on December 4, 1958 (Courtesy of the Mines and Geology Division, Jamaica)

Hynes of the Halse Hall estate in Clarendon, in 1754. Helen was the daughter of Jannett Guthrie, daughter of James Guthrie, Custos of Westmoreland, and Conciliator of the Maroons. It was Guthrie in March 1739, commanding a large body of Militia, with Lieutenant Francis Sadler, concluded the famous treaty with the Maroons who were led by Captain Cudjoe. Guthrie's daughter, Jannett, had been married to John Hynes of Westmoreland, but after he died Lieutenant Sadler took her as wife. Sadler was the half-brother of Thomas Hals, owner of Halse Hall. The last of the Jamaican extension of the Hals family died in 1743, and Sadler inherited the property and changed his name to Sadler-Hals in May 1746. He died without children and Jannett inherited the estate. She outlived her daughters from her previous marriage, Helen, and Elizabeth Hynes, and remained the owner of the estate until she died in 1775 (Chubb 1958).

The estate passed on to Thomas Beach II, one of eight children born to Thomas I. Thomas II moved to England in 1777, two years after his grandmother's death, and the estate was left in the care of the Attorneys, Messrs. Simpson, and Taylor of Kingston. Thomas II joined the army and moved up the ranks to Lieutenant Colonel. He also changed his name to De la Beche in 1790. He later married Elizabeth and the two lived on Welbeck Street, Cavendish Square where Henry Thomas was born. Thomas II ensured his estate was in proper order, and this was done through his many visits to the island. On one occasion he was accompanied by his son Henry. The Colonel was sickly, and part of his trip included a visit the therapeutic waters of the Bath mineral spring in St. Thomas-in-the-East. He, however, died there on June 1, 1801 (Chubb 1958). Henry Thomas was raised by his mother in southwest England, living first in Devonshire.

By 1812 the family moved to Dorset where the rocks abounding in fossils likely stimulated Henry's geological interests (Sharpe 1997).

1.7.1 Visit to Jamaica

Like most planters, the owners of Halse Hall were absentee proprietors who lived in Europe enjoying the returns of their estates. The day-to-day operation of the estate was delegated to attorneys and overseers (Black 1976). Henry Thomas inherited Halse Hall, and to ensure that the family estate was properly managed, he spent a year in Jamaica. On November 9, 1823, he boarded the ship *Kingston* docked in the Bristol Channel. The vessel reached Kingston, Jamaica on Sunday, December 21 where De la Beche spent Christmas, enjoying the festivities. This stop-over presented the opportunity to observe the landscape at that part of the island. New Year's Eve was spent touring Halse Hall. He noted that the lower sections of the estate were susceptible to flooding by the Rio Minho and were draped by sediments. He recorded that the Great House was located on a hill of white limestone and identified a small gravel hill to the southeast of the Halse Hall works where he found agates of good quality (Chubb 1958).

Using the income from his estate De la Beche began a series of geological expeditions across Jamaica in 1824. He toured the central parts of the island visiting places in St. Catherine, Clarendon, and Manchester. On his birthday, February 10, he began a three-day excursion down the Clarendon Plains to Milk River then up the fault-controlled Sixteen Mile Gully to the Carpenters Mountain and then southwest to Plowden Hill which afforded him a view of St. Elizabeth as far as the southern tip of the Santa Cruz Mountains and Alligator Pond Bay (Chubb 1958).

By the end of February De la Beche set off on his longest tour that lasted more than six weeks. Having gone to St. Thomas with his ailing father when he was a child an exploration of the eastern parishes was top on his itinerary. In March he explored the Kingston Harbour and cays then made his way up the Liguanea Plains and along the Hope River Valley to Middleton in St. Andrew to visit a Frenchman, Mr Castile. The proximity of Middleton to the Blue Mountain peaks allowed him to estimate their heights. The first he estimated was that of Catherine's Peak (or St. Catherine's Peak as he had noted in his diary) accurately judging a height of 5000 ft. (Catherine's Peak is surveyed at 5060 ft.) The summit was named after the sister of Edward Long (author of Long 1774a, b) of the Longville Estate of Clarendon, who was the first woman to scale the peak. On March 6, De la Beche climbed to the top and viewed the other peaks on the Grand Ridge to the east. After leaving these mountains he toured St. Mary as far as Port Maria. He visited the old run-down Fort Haldane which he described as 'dreadfully out of repairs' (Chubb 1958). He went as far as Buff Bay and Charles Town in Portland before returning to Kingston.

Towards the end of March 1824 De la Beche explored eastern Jamaica which included expeditions into the Blue Mountains. He was accompanied by five men who had similar interests, and 20 menservants. The party travelled through Guava Ridge and ascended to the peak by way of Portland Gap, scaling a 'troublesome' forested slope. He proposed a height of 7500 ft. for the Blue Mountains, stating that the old figure of 8080 ft. was overestimated. (The current surveyed height is 2256 m or 7402 ft). He later toured the eastern-most side of the island starting at Albion then ascending the Yallahs Valley noting the narrowness of the gorge above Norris. He passed the deserted works of Llandewey which are opposite a mountain with a precipitous cliff at the summit (Chubb 1958). What he observed was Mount Sinai and Judgement Cliff, the scar of a large earthquake-induced landslide that occurred in 1692. Touring eastern St. Thomas, he recalled his 1801 visit to Bath as a child. He noted that the two-acre botanical gardens were 'not in very good order' (Chubb 1965). He later traversed the Cuna Road to view the head of the Rio Grande Valley. In early April he left Bath for Pleasant Hill, but not before viewing *Carrion Crow Hill* or the John Crow

Mountains. He spent a day exploring the Rio Grande Valley, riding up to the Golden Vale estate, then on to Moore Town before returning to Port Antonio to spend the night. On leaving Port Antonio De la Beche headed west examining the rocks in the other major river valleys before heading towards home (Chubb 1958).

De la Beche made some remarkable conclusions about the rocks he observed. His visit to Savanna Point in Portland led him to note:

> I fell in with cliffs which led me to consider Black Hill an extinct volcano - the hill has a peculiar shape. (Chubb 1958)

His observation of Black Hill as a volcanic centre was indeed correct.

Maintaining the hypothesis that all the land masses across the world were connected, he managed to identify similarities of many of the local formations to those known in Europe. He noted that the white compact limestone of Jamaica occurring in the Carpenters Mountain and St. Johns Mountain is the equivalent in this section of the world to the Permian Magnesian Limestone of England or the Zechstein (Chubb 1958). In concluding he states:

> This mixture of white compact limestone, sometimes almost crystalline, with red marl and sandstone, I shall for the future call the White Limestone Formation as that rock is the predominant and characteristic feature . . .. (Chubb 1958)

De la Beche supported his observations with geological cross-sections. He made a coloured geological section from Lluidas Vale to Old Harbour where he mentions 'some light-coloured marls would appear to separate the White Limestone and St. John's trap rock near Wood Hall'. These marls would later be named the Yellow Limestone Formation (Chubb 1958). At the Newcastle area north of Kingston he figured that he had found the lower Permian red beds of Europe, and Coal Measures. The red beds occur as red conglomerate below the Coal Measures. These were his observations of the formations known today as Wag Water Conglomerate and Richmond Shale, respectively (Chubb 1958). Other cross-sections included one from Long Mountain to Catherine's Peak, the second from the Kingston plains, through Stony Hill to the shore in St. Mary, and the third from the southern Port Royal Mountains to the Blue Mountains.

He ended his excursion of eastern Jamaica on early April 8 and was somewhat fortunate to experience an earthquake on that day. Two days later, a second shock was felt, followed by a third on April 13. Over the next months he continued excursions near home, but the summer heat proved a challenge for him. De la Beche left for England in late December 1824 (Chubb 1958) and three years later his work which included a coloured geological map and some cross-sections was published (Fig. 1.5; De la Beche 1827). At that time the concept of a geological map was still relatively new. Among his published works arising from his visit, De la Beche also shared his interaction with the slave population and the resulting culture shock (De la Beche 1825).

1.7.2 Surveying Britain

In 1791 the Ordnance Trigonometrical Survey was established to prepare accurate maps of the British Empire to aid in its defence plans. This trigonometrical survey was under the Board of Ordnance, a body set up by law to regulate the manufacture of artillery for the defence forces of the Crown. De la Beche conceived the idea of using the maps prepared by the Ordnance Survey of Britain as the basis for geological study. In 1830, again using his estate income, he attempted a geological mapping of Devon, but self-financing this venture would not prove sustainable. The Jamaican estate left him at a loss, and he was forced to seek funding from the British Government to complete the research. The Ordnance Survey was still in progress, and he explored the opportunity to work in tandem with the topographic surveyors (Chubb 1958).

The first geologist to be appointed to this survey was John MacCulloch. He started as a chemist with the Board of Ordnance and was later dispatched in 1811 to Scotland to locate areas where geodetic errors could be expected,

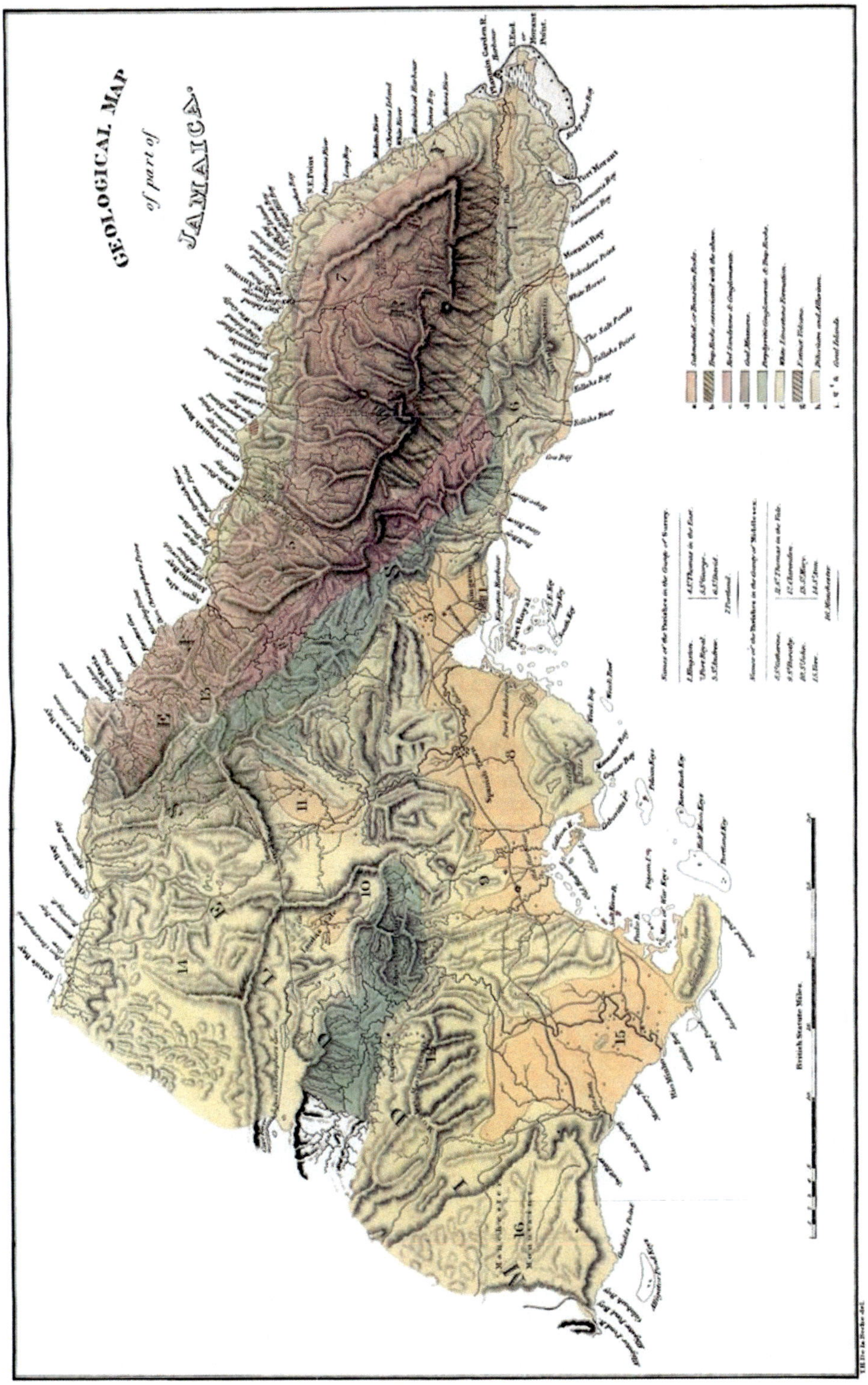

Fig. 1.5 Geological map of eastern Jamaica prepared by Henry T. De la Beche At this time, geological maps were novel and only a few countries like Great Britain and the United States had maps covering significant portion of their landscape (Adapted from De la Beche 1827)

and further to establish a reference meridian for the construction of the one-inch map (Wilson 1985). MacCulloch used the time in Scotland to observe the geological features and recorded these as partial surveys of the country. He continued with the Board of Ordnance until 1826. After this he was paid by the Treasury to complete his geological map of Scotland until his tragic death on his honeymoon in an unfortunate carriage accident in 1835 at 62 years. He had constructed a map of all of Scotland on a one inch to four miles scale, and this was published posthumously in 1836 (Wilson 1985).

The Superintendent of the Ordnance Survey, Thomas F. Colby recognized the practical usefulness of the geological survey and supported De la Beche in securing public funds for the mapping of Devon. De la Beche had single-handedly mapped over a thousand square miles of Devonshire in three years (Wilson 1985). After successfully completing this mapping, De la Beche, then the Vice-President of the Geological Society, attempted to convince the English Government to continue the survey into the neighbouring county of Cornwall, an area known for its tin deposits. The Society supported the idea of combining the geological and geographical surveys. De la Beche had completed geological maps on eight Ordnance Sheets within the period from 1832 to 1835 (Cook and Allen 1992). The Secretary of the Treasury questioned the time it would take to complete the mapping of Britain and De la Beche wisely responded that it would run concurrently with the production of the trigonometrical survey maps. Not impressed, the Treasurer cited that the mapping of Devon took three years, and at the same rate of mapping the remaining area would take 21 years if he was to work alone. They therefore considered employing two or three other qualified geologists, to reduce the time to seven years (Cook and Allen 1992). The Ordnance Geological Survey was therefore approved and established initially as a one-man department under the Board of Ordnance. This new department was a part of the Ordnance (Trigonometrical) Survey and De la Beche was appointed its only full-time employee and first Director (Wilson 1985). He would serve in this capacity until his death twenty years later.

Cornwall was next to be surveyed, commencing in 1835. That year marked the official establishment of the Geological Survey of England and Wales. De la Beche, the new Director, received a salary of £500 per annum (Chubb 1958). He managed to secure the needed funds to sustain the surveys, and it is recorded that in 1839 by way of letter De la Beche managed to convince the Government to grant him £300 to 'colour geologically' a few sheets of the Ordnance Survey. When the survey of Devon, Cornwall and Somerset was near completion he acquired approval and funding to continue into South Wales (North 1939). Surveying assistance was obtained, though inconsistently, from William Edmond Logan, copper smelter, and from Henry McLauchlan, an Ordnance surveyor (Cook and Allen 1992). Logan would later be associated with the Geological Survey of Canada.

De la Beche was also instrumental in the establishment of several other organizations. These included the Museum of Economic Geology in 1835 (later the Museum of Practical Geology and then the Geological Museum), the 'Government School of Mines and of Science applied to the Arts' (later the Royal School of Mines) and the Mining Record Office, a repository for plans of abandoned mines which later became the Mines Department of the Home Office and the technical branch of the Ministry of Mines in 1839. He served as consultant to the Admiralty on choosing suitable fuels for the ships, and to the Board of Health on the planning and sanitation of towns. He investigated the causes of explosions in coal mines and how they could be prevented. In academia he served as Secretary of the Geological Society of London from 1831 to 1832 and was elected President in 1847 serving until 1849 (Chubb 1958; North 1939, 1951).

De la Beche is remembered for his initiative to link science to the public service through the creation of the Geological Survey of Great Britain. It was from this movement that many other colonial geological surveys were established. These include the Geological Survey of India, set up in 1851, and others such as the Newfoundland Geological Survey and the Geological Survey of New South Wales. Many of the members of the British Survey served as Directors or Chief Geologists in these Survey organizations (Cook and Allen 1992). De la Beche saw his conception evolve from being the one-man geological branch within the Department of the Ordnance Survey to become the Geological Survey of Great Britain and Ireland authorized by the Geological Survey Act of 1845 (British Geological Survey 2008). By this statute the Geological Survey was established independent of the Board of Ordnance with an increased staff complement, new posts, and the Geological Branch of the Ordnance Survey of Ireland was incorporated with it (Chubb 1958). Henry Thomas De la Beche was knighted in 1842, with his ex-wife Letitia assuming the title of 'Lady De la Beche' though they had divorced back in 1826 (Chubb 1958; Thomas 2009). He was also a Knight Commander of the Danish Order of Dannebrog, and of the Belgian Order of Leopold (Sharpe 1856). He died on April 13, 1855, at the age of 59. He is remembered for establishing geology as a profession. With all being said, Dr F. J. North, Keeper of the Geology Department at the National Museum of Wales, summarizes the life of De la Beche:

> The story of his life...is full of interest and inspiration, whether we follow him as a young man making the Grand Tour; as a plantation owner in Jamaica interested in the physical and spiritual welfare of his slaves; as a pioneer in the newly emerging science of geology; as a successful overcomer of official inertia, persuading the government of his time to establish a geological branch of the Ordnance Survey [and many other organizations]; as adviser to the Admiralty...; as pioneer in [geological] enquiries... and in propaganda... and last by no means least, in his ordinary human relationships – loyal and appreciative of loyalty, sincere and intolerant to hypocrisy, as jealous for the reputation of his assistants as for his own, as pleased as the next man when praised and as resentful of criticism that he felt undeserved... but it is abundantly clear that his greatest contribution to the advance of science... lay in his success in securing, for the first time on a large scale and on a permanent basis, Government recognition and support for scientific investigation. (North 1951)

De la Beche's work in Jamaica represented a milestone where he undertook the first proper geological description of the eastern half of the island (Fig. 1.4b) (De la Beche 1827). Though not the first attempt to map geologically in the Western hemisphere, it is one of the earliest.

1.8 From Agriculture to Mining

The failure of Jamaica's sugar industry began around the 1830s or so and this triggered economic decline in the colony. Sugar was considered the business of the rich and many planters would make a great show of their riches, especially in Europe. The expression 'as rich as a West Indian planter' (Black 1976) was the common descriptor for a wealthy person. With the economic downturn some planters were desperate to save their investment and some resorted to mineral exploration (Stafford 2002). The increase in local copper exploration and mining correlates with the decline of slavery and may have been driven too by the demand for metal resulting from growing industrialization in North America and Europe. In 1841, Scottish geologist Roderick Impey Murchison who would later become the first Director-General of the Imperial Survey (De la Beche was titled 'Director') was asked by the Prime Minister, Lord Palmerston, to provide a technical opinion on the newly discovered copper deposits of Jamaica which were generating much interest and attracting large capital investments. These are the deposits that came to the fore in the mid-1830s. He wanted to compare its value with the rich ore from Cuba that was being supplied to England. Specimens of copper and what were thought to be gold and coal specimens were also sent to Murchison by a distressed planter who was optimistic that the chief geologist would prove them valuable and provide hope for his estate on the verge of collapse (Stafford 2002). De la Beche, the authority on Jamaican geology, was consulted but offered little help considering that he did not investigate the mineral potential (Stafford 2002). De la Beche's personal geological survey conducted in Jamaica (De la Beche 1827) was the key reference for the island for several decades and its author was respected for his expertise. However, scientific investigations of this nature though largely appreciated by the community of scholars were of limited value to those interested in practical geological applications such as mining.

Huge financial losses had been incurred during the short-lived mining boom of the mid-1830s—early 1840s, and the situation repeated in the 1850s before it was concluded that a detailed geological understanding of the island's metalliferous districts was necessary before committing to new investment in the development of mines. Several mines were established and were indeed productive (Scotland 1889; Zans 1955). However, technical expertise to guide further development was needed and the capitalists saw it critical to petition of the Government for the deployment of an expert geologist to provide advice on the potential of their mines. They made consistent pleas to the 'efficient and enlightened administrator' Sir Henry Barkly, the Governor of Jamaica. As a result, Barkly sent a despatch dated December 22, 1853, to the Colonial Office containing an urgent request for a complete and systematic mineralogical survey of Jamaica (Chubb 1964). Barkly, a former planter from Guyana had come to Jamaica to settle a fall-out in relations between the Colonial and Imperial Governments, having been successful in resolving a similar situation in Guyana. Despite the financial challenges facing the colony at the time, Jamaica welcomed him and his wife with 'sumptuous entertainment and ball held in the House of Assembly' (Black 1976). The capitalists, also recognizing that 'official inertia' was likely, privately arranged for a consultant mining geologist from America, Dr Isaiah Deck to visit the island (Tenney 1854). Deck spent several months in 1854 visiting about 25 mines and making rapid assessments of their economic potential.

The Imperial Government, it seems, delayed responding to Barkley, possibly recognizing the intention to use the Government resources to justify the investment in the mines. In the latter half of 1855, nearly two years after his request, Barkly received a somewhat favourable response from the Secretary of State, Sir William Molesworth. He informed him that a qualified geologist and mineralogist would be assigned to undertake a programme of geological survey of the West Indian colonies, specifically focusing on their mineral resources. However, the survey would first be conducted in Trinidad, after which other colonies including Jamaica would have the opportunity to be investigated, if they so desire (Chubb 1964). The despatch was delivered by Barkly to both houses of the Legislature and the matter was referred by the Assembly, to a Select Committee which recommended acceptance of the offer. The Committee responded that:

> . . . capital [for mining] can only be secured through confidence in the mineral character of the country. This character has yet to be established, and your Committee can conceive no means so likely to lead to a just conclusion of this most interesting and important subject as the official survey of its mineral and geological features, by an officer to be appointed by Government (Chubb 1964).

The acceptance letter of the Select Committee was sent to the Colonial Office in London where it came to the attention of senior official Henry Taylor who commented that:

> . . . it is very important that the person selected [as geologist] should be an honest man, for it is probable that large pecuniary and commercial interests will be concerned in procuring official authentication of the mineral advantages offered by particular colonies and districts (Chubb 1964)

Taylor, a long-standing civil servant, was among those who had plans to acquire a plantation in the Caribbean but was bitterly disappointed by the Emancipation bill of 1834. In 1836 Taylor was offered Governorship of northern Canada but committed to the quiet life of Civil Service in the Colonial Office where he had worked since 1823 (Anonymous 1885).

Even though Barkly's request for a survey may have prompted the programme for the

Caribbean territories, the Imperial Government was careful and systematic in their approach. The Governor of Trinidad had presented a more desperate case which served as a convenient strategy for delay. In 1855 he sent specimens of iron pyrite said to contain gold to the Colonial Office for analysis. This was accompanied by an application from a local gentleman requesting a Government Geologist to survey the mineral wealth in that island. The Trinidadian specimens were found to be useless. This however compelled Murchison, who had just been appointed Director-General of the Geological Survey of Great Britain and Ireland, to maintain the power of the Imperial Survey to determine and appoint Government geologists. He proposed that island geological surveys should be conducted by 'a really intelligent agent' to expose the fraudulent claims by mining speculators and do more in the colony beyond simply exploring for gold (Stafford 2002). Murchison's proposal was accepted by the Colonial Office and all Government geologist appointments and chemical assays and analyses for geological work were conducted by the Imperial Survey back in England. Murchison had just succeeded De la Beche who died in the same year. Murchison, exercising great perspicacity, encouraged the establishment of a geological survey entity for the West Indies instead of disjointed assessments. Like the vision of De la Beche for Britain, Murchison saw the greater benefits that a systematic survey would have on the future development of the colonies. The capitalists, however, were more interested in surveying their mines rather than a national survey. This confusion would later manifest.

1.9 Global Gold Rush

The boom in copper mining in Jamaica in the 1850s correlates temporally with gold rush events in other countries, and these global movements may have driven mineral development in Jamaica. The gold rushes were attractive to many Jamaicans who faced hard times. The California Gold Rush probably represents the beginning of this worldwide wave. The Mexicans had always found gold in California; however, mining became more lucrative when on January 24, 1848, at Sutter's Mill near Coloma, James W. Marshall noticed pieces of golden metal in the tailrace of a lumber mill he was building along the American River. He secretly tested his findings and confirmed that it was gold. News soon spread of the discovery, and this triggered a mass influx of persons to the state. By the beginning of 1849, word of the Gold Rush had spread around the world and prospectors and merchants converged from all over (Bancroft 1888). Jamaicans too sought to benefit from the American gold. Two years later in 1851, gold was discovered in New South Wales, Australia, by Edward Hammond Hargraves at a site he called Ophir near Bathurst. News of this spread and in 1852 alone 370,000 immigrants arrived in Australia (State Library of New South Wales 2014), and again, Jamaicans were in the lot. The excitement became greater on July 20, 1851, when Thomas Peters found gold dust in what is now known as Specimen Gully in Victoria. This discovery was published in the Melbourne *Argus* of September 8 triggering yet another gold rush (Goodman 1994).

In Jamaica, sugar was no longer important, the economy had collapsed, and the Imperial Government paid minimal attention to the local affairs. Further, public health was threatened by a serious outbreak of cholera. Many colonists who could afford it opted for relocation to the Australian continent citing that the old plantation economy was collapsing after emancipation and that they saw 'barbarism' growing all around. The cost of the passage was high and as such only a few black labourers were able to travel (Higman 1976). One reader of the *Falmouth Post* asked in a May 1852 issue for the 'shortest, safest and cheapest passage to Australia' (Higman 1976). The editor reminded him that the Jamaicans who went to California returned broke. It was said that a person could work for six months in the goldfields and find gold valued at the equivalent of six year's pay back home. However, on average, only half of the gold seekers made gains, and this was only modest, after considering the expenses. Most of them, especially those who came later, which included the Jamaicans, made little or ended up in losses. The editor reminded that Jamaica needed men 'who work with the head'. Other writers rebuked the potential emigrants on their willingness to work hard overseas but not in their homeland. They blasted that they are blocked by their pride and view cultivating their native soil 'degrading employment'. In a June 1852 issue of the Kingston *Colonial Standard* an optimistic letter from Rev. John Gibson, a Presbyterian missionary, was published. He had served in Jamaica for several years but left for work in New South Wales (Higman 1976). His views for Australia were positive and very encouraging. Unfortunately, the opposite was his opinion for Jamaica:

> The destiny of this colony [Australia] is quite changed by this great discovery. Jamaica is poor compared to it. How well the black people and others from your country would do here! I wish we had one or two thousand. We expect thousands of emigrants for our gold. (Higman 1976)

Response to his encouragement was immediate. Within a week nearly one hundred people left for Panama to board a ship crossing the Pacific for the 'land of milk and honey'. Further, a group of Kingston 'gentlemen' formed a company to charter a vessel to transport 'respectable persons' to Australia. By the end of June 1852, the first ship to leave Kingston for Australia was already booked to capacity. Passage cost was £40, and departure was scheduled for early September. Those who were unwilling to wait travelled to Australia through London, New York, or Panama (Higman 1976).

1.10 Survey Needed!

The mining investors in Jamaica were desperate for Government assistance in investigating their mines. The Imperial Government however recognized that in the interest of the Crown a complete survey of the colony at that level like that conducted by De la Beche in England is as far as they should go. The Caribbean colonies were approached about jointly financing their geological survey activity with the Imperial Government. Murchison proposed that the salary of the new geologist should be £300 with other allowances for personal expenses. There would also be an assistant geologist. The new Secretary of State for the Colonies, Henry Labouchere, agreed with Murchison but clearly warned that the incumbent should be 'immune to pecuniary interests of a commercial kind which may be involved in procuring favourable official reports of the advantages offered by particular Colonies or Districts' (Stafford 2002).

References

Anonymous (1852) Gold in Jamaica. Sci Am 7(45):355–355

Anonymous (1885) Sir Henry Taylor, an autobiography of Henry Taylor in two volumes, with portrait. Harper and Brothers, New York, New York Times, 19 Apr 1885

Anonymous (2011) Jamaica Parish reference. http://prestwidge.com/river/jamaicaparishes.html. Accessed 14 Nov 2011

Bancroft HH (1888) History of California, Volume 23: 1848–1859. The History Company, San Francisco

Barham H (1718) A letter . . . giving a relation of a fiery Meteor seen by him in Jamaica, to strike into the Earth; with remarks on the Weather, Earthquakes, etc. on that Island. Philo Trans Royal Soc 30(2):837–838

Bellin JN (1758) Carte Particuliere De L'Isle De La Jamaique Dressee au Depost des Cartes Plans et Journaux de la Marine (Scale 1:165,000 estim.). Depot De La Marine, Paris

Black CV (1976) History of Jamaica. Collins Clear-Type Press, Great Britain

British Geological Survey (2008) Our past. British Geological Survey, Natural Environment Research Council. http://www.bgs.ac.uk/about/ourPast.html. Accessed 24 Sept 2009

Browne P (1789) The civil and natural history of Jamaica. B. White and Sons, London

Bryan P (1992) Spanish Jamaica. Caribbean Quarterly 38(2/3):21–31. JSTOR. http://www.jstor.org/stable/40654177. Accessed 20 Jan 2023

Charles II (1661) A proclamation for the encouraging of planters in his majesties island of Jamaica in the West Indies (Whitehall, December 14, 1661). John Bill and Christopher Barker (Printers to the King's Most Excellent Majesty), London

Chubb LJ (1958) Sir Henry Thomas De la Beche. J Geol Soc Jam (Geonotes) 1:1–26

Chubb LJ (1964) Lucas Barrett, A Biography. J Geol Soc Jam (Geonotes) 6:3–45

Chubb LJ (1965) The Bath of St. Thomas the apostle in the first quarter of the nineteenth century. J Geol Soc Jam (Geonotes) 7:37–38

Cook PJ, Allen PM (1992) The example of the British Geological Survey: past, present and future. British Geological Survey, Nottingham

Cromwell O (1655) By the Protector, a proclamation giving encouragement to such as shall transplant themselves to Jamaica (Whitehall, October 10, 1655). Henry Hills and John Fields (Printers to His Highness), London

Cundall F, Pietersz JL (1919) Jamaica under the Spaniards. Institute of Jamaica, Kingston

De la Beche HT (1825) Notes on the present condition of the Negroes in Jamaica. Cadwell, London

De la Beche HT (1827) Remarks on the geology of Jamaica. Trans Geol Soc Lond 2(2):143–194

Edward AH (1942) The possibilities of development of the mineral resources of Jamaica. Geological Survey Department, Kingston

Gardner WJ (1971) A history of Jamaica, from its discover by Christopher Columbus to the Year 1872 (Third Impression). Frank Cass & Co., Ltd., London

Goodman D (1994) Gold-seeking: Victoria and California in the 1850s. Stanford University Press, Stanford

Gookin D (1656) To all persons whom these may concern, in the several towns, and plantations of the United Colonies in New England (March 25, 1656). Daniel Gookin, Boston

Higman B (1976) Jamaicans in the Australian gold rushes. Jam J 10(2,3,4):38–45

Hose HR (1950) The geology and mineral resources of Jamaica. Col Geol Min Res 1(1):26–27

Jamaica National Heritage Trust (2011) The Tainos. http://www.jnht.com/download/the_tainos.pdf. Accessed 19 Aug 2021

Lee JW (1985) A pre-columbian gold artifact from Jamaica. In: Proceedings of the Tenth International Congress for the Study of the Pre-Columbian Cultures of the Lesser Antilles, Université de Montréal, Montreal, pp 343–345

Lee JW (2006) Jamaican Redware. In: Atkinson L-G (ed) The Earliest Inhabitant: the dynamics of the Jamaican Taino. University of the West Indies Press, Kingston, pp 153–160

Long E (1774a) History of Jamaica, or general survey of the ancient and modern state of that Island, with reflections on its situation, settlement, inhabitants, climate, products, commerce, laws, and government, Volume II. T Lowndes, London

Long E (1774b) History of Jamaica, or general survey of the ancient and modern state of that Island, with reflections on its situation, settlement, inhabitants, climate, products, commerce, laws, and government, Volume III. T Lowndes, London

Murphy J (2019) TAINO FIND - Israeli Miner turns over artefacts, remains found in Hellshire to Heritage Trust. The Gleaner, 07 Aug 2019, https://jamaica-gleaner.com/article/lead-stories/20190807/taino-find-israeli-miner-turns-over-artefacts-remains-found-hellshire. Accessed 30 Sept 2023

North FJ (1939) H. T. De la Beche: geologist and business man. Nature 143:254–255

North FJ (1951) Sir H. T. De la Beche: his Contributions to the advancement of Science, and the circumstances in which they were made. Bull British Soc Hist Sci 1(5):111

Padrón FM (2003) Spanish Jamaica (Translated by Bryan PE, Gronow MJ and Moral FO). Ian Randle Publishers, Kingston

Prestwidge W (2008) Jamaican Parish reference. http://web.archive.org/web/20110515000000*/, http://prestwidge.com/river/jamaicanparishes.html. Accessed 6 Dec 2021 (Site is now archived)

Renny R (1807) An history of Jamaica. J. Cawthorn, London

Rodney-Harrack N (2006) Jamaican Taino Pottery. In: Atkinson L-G (ed) The earliest inhabitant: the dynamics of the Jamaican Taino. University of the West Indies Press, Kingston, pp 146–152

Roobol JM, Lee JW (2006) Petrography and source of some Arawak Rock Artefacts from Jamaica. In: Atkinson L-G (ed) The earliest inhabitant: the dynamics of the Jamaican Taino. University of the West Indies Press, Kingston, pp 131–145

Roxburgh TL, Ford JC (1901) Handbook of Jamaica for 1901, 21st year of publication. Government Printing Office, Kingston

Sawkins JG (1869) Reports on the geology of Jamaica, or Part II of the West Indian Survey, with contributions from G. P. Wall, Lucas Barrett, Arthur Lennox and C. B. Brown, and an appendix by R. Etheridge. (Memoirs of the Geological Survey), Longmans, Green, and Co, London

Sawkins JG, Brown CB (1865) Geological map of Jamaica—scale: one mile to ¼ of an inch. Vincent Brook, Day & Son Lith., London

Scotland H (1889) On mining in Jamaica. Mortimer C. DeSouza, Kingston

Sharpe D (1856) Henry Thomas De la Beche—Obituary. Q J Geol Soc 12:xxxiv–xxxvii

Sharpe T (1997) The archive of H. T. De la Beche (1796–1855) in the National Museum of Wales, Cardiff, UK. J Geol Soc Jam 32:29–35

Siriunas JM (1993) Draft report no. 27 inventory of metallic mineral occurrences in Jamaica. Geological Survey Division, Kingston

Stafford RA (2002) Scientist of Empire: Sir Roderick Murchison, Scientific exploration and Victorian imperialism. Cambridge University Press, London

State Library of New South Wales (2014) Gold. http://www.sl.nsw.gov.au/discover_collections/history_nation/gold/index.html. Accessed 20 Apr 2014

Steers JA (1940) The Cays and the Palisadoes, Port Royal, Jamaica. Geographical Rev 30(2):279–296

Tenney WJ (1854) Mineral lands in the Island of Jamaica. Mineral Mag, New York 3:200–204

Thomas J (2009) Joan Thomas. http://joanthomas.ca/for-book-clubs/sources/henry-de-la-beche-1795-1855. Accessed 24 Sept 2009

Wilson HE (1985) Down to Earth: one hundred and fifty years of British Geological Survey. Scottish Academic Press Ltd., Edinburgh

Zans VA (1955) Geology and mining in Jamaica. The Gleaner, Kingston, 10 May 1955, p 19

2 Geological Survey for Jamaica

Abstract

With Jamaica facing economic challenges from the decline of its once prosperous sugar industry, the colonists turned to mining as a potential income earner. However, many mines failed for lack of adequate geological characterization of the mineral deposits. The miners sought this technical assistance from the Government, in addition to official endorsement that their mines were worth investment by English capitalists. The first Government-commissioned Geological Survey entity was established to investigate the mines, but more importantly, the general geological features of Jamaica. Over a period of ten years the organization would cycle through five geologists who worked to prepare a voluminous report and corresponding maps of the country. The maps, however, were not prepared in great detail, limited by the absence of detailed topographic maps as a base. After the geological survey activity had ended, there were numerous calls for updates and more detailed studies. Further, the island also became susceptible to quacks in mining and geology. Despite these, the reports and maps of that first Government Geological Survey would persist as the standard official references on Jamaican geology for the next 80 years.

2.1 The Call for a Survey

Acting on the advice of Director-General of the Geological Survey in Britain, Rodrick Murchison, the English Government in 1855, decided that the geological survey programme for the Caribbean territories should begin with Trinidad. This activity would be a 'general survey of the economic geology' (Wall and Sawkins 1860). Governor Henry Barkly of Jamaica, who had initiated the request for a survey of the mineral potential of Jamaica in 1853, despite being notified of the preference for Trinidad, still sought for his colony to be prioritized:

> It's great importance and utility cannot be better described than in the words of the Committee (a Select Committee established by the local legislators to examine Britain's offer for a geological survey), and it is certainly no small guarantee of the good faith of the mining speculators of Jamaica, that those who represent their interests have displayed so much anxiety to have a competent officer appointed, to investigate the extent and reality of the mineral developments of the Island.
>
> I observe that Her Majesty's Government determined upon the employment of such an officer primarily in the case of the Colony of Trinidad, but whilst far from desirous of raising any question as to the propriety of his first proceeding thither, I may venture, perhaps, as in duty bound, to prefer the interests of the Colony committed to my charge, to call to mind, that so far back as in my despatch of 22nd December 1853

S. Bhalai, *The History of Mining and Geological Surveys in Jamaica*, Historical Geography and Geosciences,
https://doi.org/10.1007/978-3-031-42604-9_2

> (No. 32), I urged that a complete and systematic mineralogical survey of Jamaica should be authorized.
>
> —Henry Barkly, King's House, November 26, 1855,
>
> (Wall and Sawkins 1860:208)

The Trinidad surveys began officially in 1856 and were titled *Part One of the West India Geological Survey*. Jamaica, Grenada, Barbados, St. Kitts, and the Virgin Islands agreed to finance their surveying activities after Trinidad was completed; however, Antigua and British Guiana declined participation. George Parkes Wall was appointed as the lead geologist and James Gay Sawkins then 50 years old, as assistant geologist (Anonymous 1912). The surveyors were chosen on recommendation of Sir Roderick Murchison. Wall, 26 years old, was one of the earliest students of the Royal School of Mines. A year before his appointment in 1855, he received the Associateship of the College of Mining. He gained practical experience in the mines at Freiberg, Germany, until Murchison presented the offer of directorship (Anonymous 1912).

Sawkins was born in England and had learnt the art of painting in Italy. He however, lived in the USA as an arts professor (specializing in drawing) at a Virginian college (Compton 2006). It seems that Sawkins gained most of his mining expertise on the job. In 1830 he moved to Mexico where he continued as a painter and occasionally evaluated mines for the companies operating there. By 1835 he was in Cuba where he lived and worked for ten years as an artist and in the mines (Sorby 1879). Sawkins was in Peru during the 1849 announcement of gold in California. He relocated to America in an exploration party deployed by a Peruvian mining company. His company, like many others at the time, failed and he relocated to the east of the Sandwich Islands in 1854, and then to Australia where he was engaged in the inspection of mining properties and in mineral exploration programmes of west and northwest Queensland (Sorby 1879). In November 1855 Sawkins was elected Fellow of the Geological Society of London with his proposers being prominent persons in the fraternity (Chubb 1964). Despite his professional affiliation his published works were few, comprising short papers on Tonga and Jamaica, and a lengthier manuscript in 1871 that described his findings from a mapping expedition in British Guiana (Sorby 1879). After Australia, Sawkins went to Jamaica where he was working as a mining engineer until 1856 when was called to assist Wall in the Trinidad surveys. His experience working in the mines in Mexico and the Caribbean, and his artistic talents were considered useful for him to offer 'efficient assistance'. He was also known personally to Murchison (Wall and Sawkins 1860; Chubb 1964).

Murchison, in his first Annual Report as Director-General commended the work of the geologists and reinforced the usefulness of the programme (Stafford 2002). The geologists however, focused on mapping the rock types rather than 'undertake a general survey of the economic geology' for which the programme was established. The colonists were dissatisfied and soon the Legislature restricted the Trinidad Survey to a rapid assessment of the economic minerals. Trinidad contributed to the costs and was therefore able to determine the scope. Despite the shift, Murchison recommended that the surveyors should be allowed to study Tobago and the coast of Venezuela to correlate the rock formations found on Trinidad with the continent. The Colonial Office agreed, and permitted the visit to Venezuela but rejected the survey of Tobago. The final report of the Trinidad survey was completed in December 1858 but published two years later (Wall and Sawkins 1860).

The work of the geologists was beneficial to understanding the mineral potential of the island. Further, they had confirmed previous assessments such as that made by Henry De la Beche on the steam potential of a lignite seam. The surveyors reexamined the deposit and found it to be useable, as boiler technology had improved to accept low-grade fuels. The Imperial Survey, impressed with the work, encouraged the Colonial Office to publish the report of the survey as a Memoir of the Geological Survey of Great Britain. The results were also presented to the Geological Society of London. The quality of the assessment was encouraging for the survey of the other colonies. Immediately after Trinidad the surveyors were deployed on a short mission to examine the mineral resources, including marble, on Bay Islands, north of Honduras. Under the Clayton-Bulwer Treaty of 1850, the USA stipulated that England should vacate the islands because of their strategic location near one of the proposed interoceanic canal routes. Lord Palmerston, British Foreign Secretary, wanted to ensure that there were no minerals of commercial importance worthy to validate retention or defer surrender of these islands (Stafford 2002).

2.2 Geologists for Jamaica

Acting Governor of Jamaica, Lieutenant General Edward Bell who replaced Barkly in 1856, received a Memorandum in October from some of the mining companies operating in Jamaica. They expressed their impatience emphasizing the lengthy delay in the arrival of a geologist to conduct 'the complete and systematic mineralogical survey of Jamaica'. It stated that several promising mines had failed, and new mineral discoveries remained undeveloped due to the lack in public confidence. Further, the miners made it clear that public confidence could only be had if there was an authoritative report sanctioned by the Government (Chubb 1964). The Memorandum was forwarded to London, and Henry Taylor of the Colonial Office recognizing that the miners sought Government endorsement responded:

> We are wrong in giving Government guidance to mining speculation. Any of the joint stock companies concerned could easily furnish the funds for a scientific survey and could go to Sir Roderick Murchison and get a geologist. What they are eager for is not so much scientific survey as Government sanction, and they will take that sanction to market and sell it high . . .. (Chubb 1964: 11)

The Trinidad surveys ended in 1858 and the Imperial Survey began planning for Part Two, the mapping scheme for Jamaica. Some of the administrators were hesitant and questioned the relevance of the survey and whether it should be financed. Murchison reminded them that little was known about Jamaica and that the activity would be useful. The survey activity faced another challenge where George Wall had resigned. It was established by the Imperial Survey that general mineral surveys of a colony should be the mandate of the colonial surveys, and the chief geologist ideally a graduate of the Royal School of Mines. Sawkins was therefore ineligible to head the programme but would still be retained. However, the low salary and limited budget available made Murchison reluctant in approaching the School of Mines graduates or even those within his own Imperial Survey. The Secretary of State for the Colonies consulted with John Phillips, Geology Professor at Oxford University and Trinity College, who recommended the acting curator of the Woodward Museum at Cambridge University, a 22-year-old geologist and naturalist Lucas Barrett (Stafford 2002).

Barrett was schooled in Royston and developed interests in palaeontology, having visited the chalk pits to collect fossils. While visiting relatives in Cambridge he met Adam Sedgwick, Woodwardian Professor of Geology at the University of Cambridge, the oldest school of geology in Britain. He also became friends with Samuel Woodward who became his mentor. Woodward was the leading expert on fossil shells and held a prominent position in the Department of Geology and Mineralogy at the British Museum. In 1853 Barrett spent a year in Germany studying the language and chemistry. He also made a walking tour of Bavaria, studying the local geology. In 1855 he was elected Fellow of the Geological Society of London but not having reached seventeen and a half years old, he was ineligible. However, with his proposers being prominent members such as Sedgwick, he came highly recommended. At eighteen years Barrett was involved in a dredging expedition beyond the Arctic Circle, Greenland, Spain, among other places. By the end of 1855 he was chosen by Sedgwick to work as Curator of the Woodwardian Museum at Cambridge. He spent three years at there and with the help and guidance of Sedgwick, he produced a geological map of Cambridgeshire. By then he had contributed eleven academic works to scientific journals. In 1856, Sedgwick, nearly sixty-eight years entrusted Barrett to deliver most of his lectures in Geology at Cambridge University. He however did not have professional qualification in geology, and upon encouragement from Sedgwick he registered at Trinity College in Dublin in January 1859 but attended school for only forty days; he never graduated (Chubb 1964).

Sir Edward Bulwer-Lytton, Secretary of State for the Colonies consulted with John Phillips on a suitable candidate for the head of the colonial Survey, and as a result an offer was made to Barrett early in 1859. Barrett was not officially appointed at Cambridge and therefore freely accepted the offer. A few days before leaving for Jamaica, he got married and with his new bride Alice Maria, set sail in March 1859. Both Barrett and Sawkins met in Jamaica early in April 1859 (Fig. 2.1a, b) and *Part II of the West India Geological Survey* began. The year 1859 marks the establishment date of the Jamaica Geological Survey; the first Government-commissioned geological survey activity for the island. The Imperial Treasury was responsible for salaries whereas allowances for personal and travelling expenses were paid from the Island Treasury (Chubb 1964; Ancestry 2021).

2.2.1 Expeditions

The geologists immediately started working using the James Robertson-drawn one-inch-to-the-mile topographic maps as their base map for navigating and recording observations. The collection, prepared from actual trigonometric surveys, comprised twenty sheets each covering a parish; at that time there were twenty-one parishes (Fig. 2.2). The surveyors first worked together mapping St. Thomas-in-the-East at that region between the Morant and Plantain Garden Rivers. This took them over five months, completing it in September 1859. Here they discovered the beds of fossil marine molluscs at Bowden and limestone outcrops in the Plantain Garden River containing the fossils *Inoceramus*, *Hippurites* and *Nerinea*. These Bowden beds were dated as Miocene and the *Hippurite*-bearing beds as Cretaceous and not Palaeozoic as De La Beche had thought. Barrett's findings were read at the February 1860 meeting of the Geological Society of London (Barrett 1860). In it he described a thick bed of grey limestone crossing the Grand Ridge of the Blue Mountains

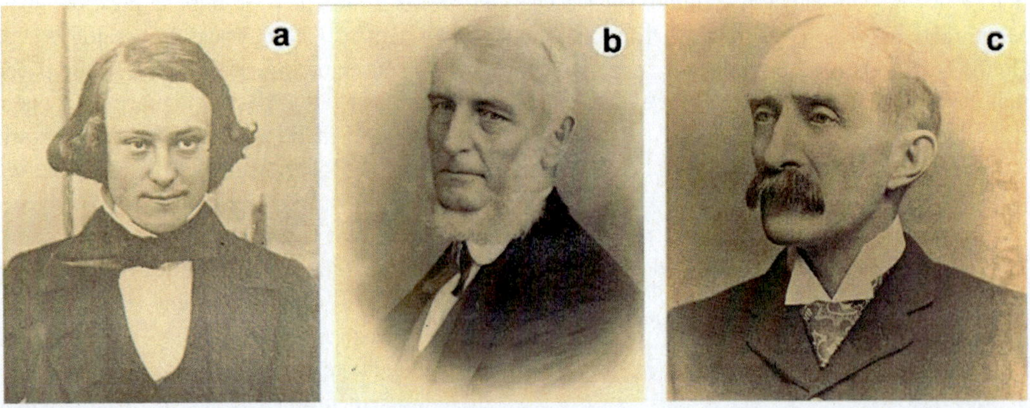

Fig. 2.1 Geologists of the first Government-commissioned Geological Survey of Jamaica in 1859. **a**—Lucas Barrett (first Director), **b**—James Gay Sawkins, and **c**—Charles Barrington Brown (Courtesy of the Mines and Geology Division, Jamaica)

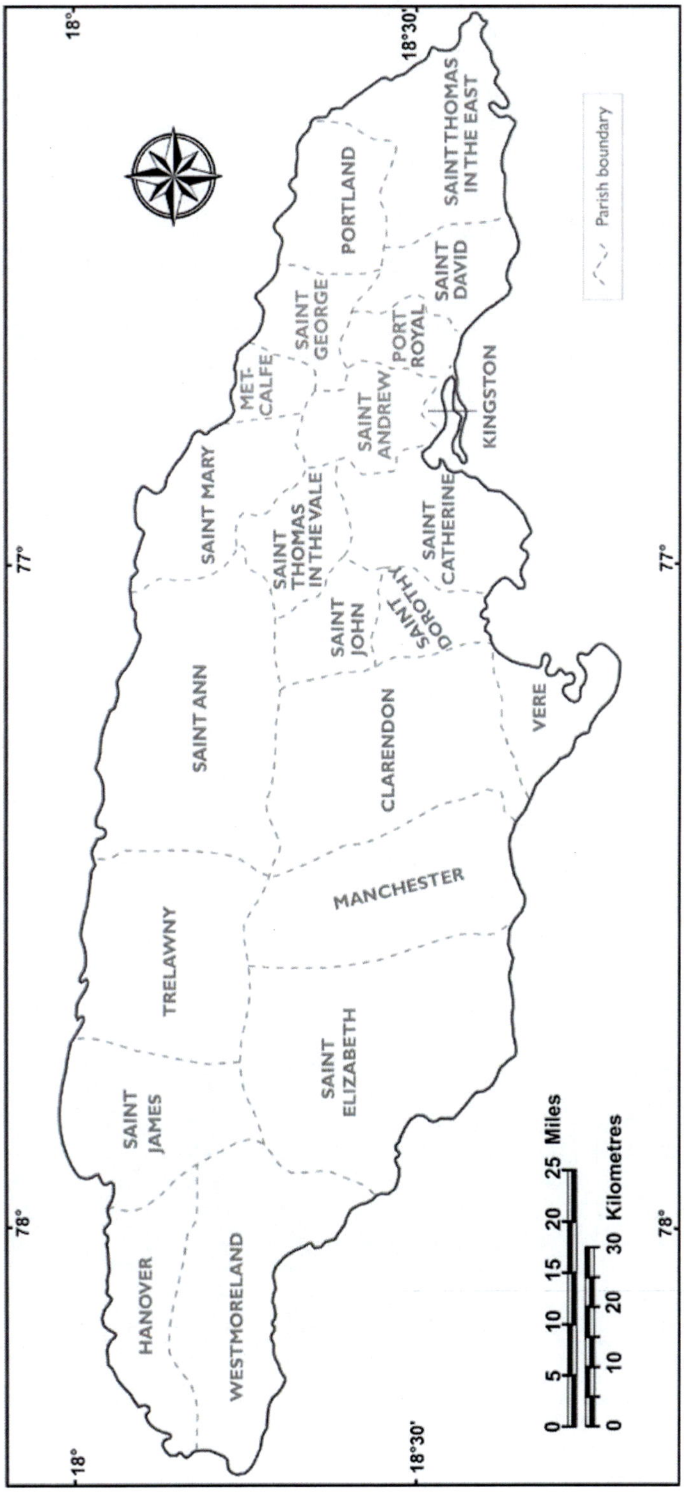

Fig. 2.2 Parishes in Jamaica (1841–1865) (Prepared from Prestwidge 2008)

at 2,500 ft. elevation having *Hippurites* and *Ventriculities* which suggests that this region was once beneath the sea (Chubb 1964).

After the survey of the parish of St. Thomas-in-the-East, the two geologists worked separately in Portland parish in October 1859. Sawkins, despite having mineral exploration experience, was assigned to survey the general geology while Barrett, a stratigrapher, was assigned, by special request of the Imperial Government, to investigate the copper mineralization in that parish (Chubb 1964). Since Spanish times Portland had been an important copper mining district where many small mines had been developed which also produced some gold (Fig. 2.3). Like the miners, the Government wanted to know about the mineral deposits. However, out of caution, Barrett was tasked this role rather than Sawkins who had just a few years before, consulted in a private capacity to evaluate the mines. Barrett's geological experience would not be attractive to the miners and would minimize the likelihood of him being approached to work as a consultant, and hence distracted from the scope of the survey. Individual reports were submitted to Governor Darling by December 1859. Barrett surveyed the deposits of copper and manganese and seams of lignite which he referred to as the 'Coal Measures' of Portland (Barrett and Sawkins 1869).

In updating Governor Darling about the first nine months of surveys, Barrett projected that a general geological survey of Jamaica would take about six years based on the size and ruggedness of the eastern regions. The establishment of a geological museum at Spanish Town, then the capital of Jamaica was also proposed. Barrett had collected rock specimens from St. Thomas-in-the-East along with common minerals and fossils from the Tertiary and Cretaceous formations. Sawkins had collected specimens of marbles, flagstones, and minerals which he

Fig. 2.3 Enlargement of northeast section of the 1865 Geological Map of Jamaica (Sawkins and Brown 1865) showing the copper mines (dots) in and around the then parish of Portland, north-eastern Jamaica. Also shown is the Spanish Mine and other mines (cluster in the west; left side of map) in St. George parish which were operated by the Spaniards (modified from Sawkins and Brown 1865)

displayed at his office in Port Antonio for 'the instruction of the inhabitants' to whom he would explain their economic potential (Barrett and Sawkins 1869). The geologists were required to prepare two collections of specimens; one for the colony and the other to be lodged in England (Wall and Sawkins 1860). In the first few years the geologists focused on eastern Jamaica particularly because it was one of the major metalliferous districts in the island. The geological reports were done by both geologists, and the poetic descriptions of the natural environment are attributed to the artistic talents of Sawkins (Chubb 1964).

2.2.2 Challenges

Within a year of working together, a disharmonious relationship developed between the two geologists which Chubb (1964) attributed to the preference given to Barrett to lead the Survey, despite not qualified by the Royal School of Mines nor experienced in surveys for economic purposes. The matter was escalated to the Governor who referred the matter to the Colonial Secretary Henry Pelham-Clinton who further consulted with Murchison at the Imperial Survey. Pelham-Clinton recognized that the issue rested with Sawkins and was willing to remove him. He however sought Murchison's input. Murchison however, advised that both geologists brought specific skills to the work of the Survey. The Governor was given the option to have one of the geologists dispatched to another island to conduct surveying activities there and a substitute geologist provided as a replacement. The immediate option was to send Sawkins to Anguilla to examine the phosphate deposits at Sombrero (Chubb 1964).

Murchison had long pressed for the exploration of an independent source of fertilizer to substitute and further liberate from the monopoly of Peru. The guano deposits off the southwest African coast were exhausted and the Americans had discovered a rich phosphate deposit on a small islet in the West Indies. Anguilla was thought to have similar potential. While in London, Sir Hercules Robinson, Governor of St. Kitts presented to Murchison rich specimens of phosphate which led to him encouraging investigation of the deposit through the Royal Geographical Society, the Royal Agricultural Society, and the Colonial Office. He stated that the investigation would be of interest not only to the geographer and geologist but also to merchants and ship-owners (Stafford 2002). Governor Darling however informed the Secretary of State that if Sawkins went to Anguilla it was likely that he would not return to Jamaica as he had received an offer of employment in Australia. Sawkins, it seems, was not sent to Anguilla at the time proposed as it was several years later in January 1865 that he prepared the phosphate assessment report dispatching it from Sombrero (Sawkins 1869a). The Australian job also did not work out. Having no other options, it seems that both geologists found a way to work together (Chubb 1964).

During the period of discord Barrett had received through the Governor, a communication from the Secretary of State containing a copy of a petition sent from mining companies in Jamaica to the Colonial Office. The 'large pecuniary and commercial interests' mentioned by Henry Taylor of the Colonial Office had expressed their dissatisfaction with the surveys. The petition asserted that the Survey geologists were engaged in a surface survey and an analysis of soils in other parts of the island that were deemed non-metalliferous. The Colonial Secretary sympathising with the miners, wanted to know 'whether a mineral survey could conveniently take precedence of a surface survey' (Chubb 1964). Barrett's response dated May 2, 1860, explained that an analysis of the soils would be useful for agricultural purposes, but indicated that no such analysis was being made. He further explained that the six months from October 1859 and April 1860 were spent surveying Portland, the parish deemed to be one of the richest metalliferous districts in the colony. He gave details that the survey had extended through a thickness of 7,800 ft. of rock, noting that the deepest shaft driven in Jamaica had only reached a depth of 390 ft. Barrett made it clear that careful examination of the geological structure was a necessity before a report of the mineral veins in the area could be made. He stressed that reporting on a mineral vein without prior geological knowledge of the whole area would mislead the English capitalists (Chubb 1964). Murchison was requested to validate Barrett's remarks. Barrett had informed Murchison that an investigation of the copper deposits was in progress, and the findings from such would be presented as one report and not specific to the mines assessed. His explanations were endorsed by Murchison who further added that establishing mining details is not the duty of the Geological Survey. He reinforced that Barrett cannot be requested to comment on the value of the individual mineral deposits to reassure the public and enhance the confidence in mining speculators (Chubb 1964).

2.2.3 Promoting Jamaica

Towards the end of 1861 the surveyors assembled a collection of metallic ores and other minerals for display in London. This was at the great International Exhibition of May to November 1862, where Jamaica would be showcased. Copies of the new Jamaica geological maps were made along with some geological cross-sections, diagrams, and other illustrations. The Secretary of State and the Privy Council of Jamaica approved and granted Barrett six months leave of absence allowing him to represent the colony at the exhibition as a Commissioner of Jamaica. Barrett and his wife departed for England on March 12, 1862 (Barrett 1869). The International Exhibition, sponsored by the Royal Society of Arts, Manufactures and Trade, was held beside the gardens of the Royal Horticultural Society, South Kensington, London. Twenty-eight thousand exhibitors from 36 countries displayed items of industry, arts and technology. Over 6 million visitors attended. The Jamaican exhibits included large specimens of minerals, metallic ore concentrate such as those of copper and manganese, building stones and a variety of marbles, all on display in the Jamaica Court. The exhibits caught the attention of the visitors and not surprisingly it was awarded two medals (Chubb 1964; Mitchell 2016). It was also at this exhibition that local distillers J. Wray and Nephew won three gold medals for their ten-, 15-, and 25-year-old rums (Lowe 1892; Jamaica Gleaner 2002).

Barrett used the visit to England to get expert assistance in identifying and correlating his Jamaican rocks, minerals, and fossils with those found in England. Among the specimens were rudist bivalves from Back Rio Grande, Portland, and molluscs and corals from Bowden, St. Thomas. There were also some foraminifers from the Cretaceous rocks and other rocks below the White Limestone. Woodward described and sketched the rudist and named it *Barrettia* in honour of its discoverer. Barrett also took the opportunity to scale the parish geological maps of St. Thomas-in-the-East, Portland, St. George, Metcalfe, and St. David, and to have the reduced maps engraved in stone and copies printed and coloured for use in the public offices of both Jamaica and England. Topographic equipment was procured for use during the surveying activities. He also obtained diving equipment for offshore surveys at Port Royal, as Sawkins was already mapping Port Royal parish. Barrett returned to Jamaica in October (Chubb 1964). He wanted to stay longer but his request was denied, and further, he was given half pay for the time in England as his visit was not entirely on official business but included other activities such as leading a field excursion to the coastal town of Hunstanton for a geological association based in Cambridge (Mitchell 2016).

2.2.4 Changes

In November 1862 Barrett submitted his Annual Report to Edward John Eyre, the new Acting Governor. He briefly commented on the assignment in England and stated that the survey of Port Royal parish by Sawkins was completed. He also stated that the Jamaican mineral specimens left with Lewis Hoffman, the Island

Chemist were analysed and proved to be very rich in metals. This included a specimen of native copper. He added that the work started on the proposed geological museum would be continued, noting that specimens were rapidly accumulating (Barrett 1869).

However, during Barrett's absence and with the retirement of Governor Darling, the 'pecuniary and commercial interests' took the opportunity to express their dissatisfaction about the survey with the new governor. Governor Darling, through correspondences with the Colonial Secretary had become familiar with the mandate of the geological survey. Eyre initially did not meet Barrett, but it seems was unfavourably briefed about the Director by the investors and their supporters in the Legislature. Barrett was summoned before a Committee on December 18 to elaborate on the status of work and to answer the accusations laid against him. He was however, determined to test his new diving suit. On December 19, Barrett, along with his boat crew and personal servants, went to Port Royal to examine the seafloor at the cays to the south. While he was supposed to be underwater, the crew reportedly saw Barrett floating on the surface at some distance from the boat. While still pumping, they retrieved him and removed his helmet, but he appeared dead. They rushed ashore, but by the time they reached the beach there was no sign of recovery (Chubb 1964).

An inquest was convened to ascertain the cause of death. They concluded that water entered the air-tube, breaching the joint, which was not tightly screwed, so that Barrett drowned. However, no water was found in the suit and the internal air pressure would have prevented water from seeping in through any loose joint but instead, air would escape. Nearly one hundred years later a medical expert concluded that Barrett died of pulmonary air embolism (Chubb 1964). Barrett died at 25 years of age and was buried at the St. Andrew Parish Church at Half-Way-Tree. A memorial shield sculpted from local Serge Island marble from St. Thomas was mounted in the corner of the north transept. It bears the words:

> In Memory of
> LUCAS BARRETT Esqr. F.G.S., F.L.S.
> Director of the West India Geological Survey
> who was drowned near Port Royal
> Decr. 19$^{\underline{th}}$ 1862.

Curiously, the words 'was drowned' rather than simply 'drown' were chosen. Barrett's obituary was read by Professor Andrew Crombie Ramsay, President of the Geological Society of London in 1864 (Ramsay 1864) and later, the successor to Murchison at the Imperial Geological Survey in 1871. He recounted the professional legacy of Barrett and sorrowfully shared that he did not live to see his only child Arthur who was born in Cambridge in January 1863, only a few weeks after his father's death (Wilson 1985; Ancestry 2021).

2.2.5 The Work Continues

With Barrett's death the post of Director of the Jamaica Survey became vacant and Sawkins applied for the job shortly after. He was promoted and was soon joined by George Wall who was on visit to Jamaica for a few months (Anonymous 1912). After the Trinidad surveys Wall had returned to Europe. On the Jamaica survey he was assigned to work in Western Metcalfe parish. He was attacked by an ex-slave wielding a cutlass and so for the rest of the time he worked alongside Sawkins in mapping St. Mary and St. Thomas-in-the-Vale (Sawkins 1866). The three reports of the work in these parishes were submitted in 1863. Wall was compelled to return to England to take over his family's steel manufacturing business (Anonymous 1912). Arthur Lennox, on recommendation from Murchison, replaced him. Lennox was a former student of the Royal School of Mines and the son of an old friend. He however retired early for failing health (Sawkins 1869b; Stafford 2002). He had surveyed St. Catherine in 1864. In that year Sawkins mapped the smaller parishes of St. John, St. Dorothy, and Vere before being deployed to Anguilla. By the end of that year, the 25-year-old Canadian-born, Charles Barrington

Brown, graduate of Harvard University and of the Royal School of Mines was appointed (Fig. 2.1c). The hiring was done through one of the recognized authorities in Geology, Sir Charles Lyell who was a close friend of Brown's father, who was a fellow geologist (Harker 1917). Brown surveyed Manchester, St. Elizabeth, Westmoreland, and Hanover between 1865 and 1866 and also remapped St. George and eastern Metcalfe which were previously surveyed by Barrett. This was the last area to be surveyed. At the same time, Sawkins surveyed Clarendon, St. Ann, Trelawny and St. James. In 1864 Sawkins went to Anguilla where he highlighted a thriving guano industry since the discovery of that resource in 1857 (Sawkins 1869a).

The Jamaican Legislature, on two occasions, suspended the annual appropriations of the the Geological Survey, but Murchison strongly objected. The Government only allowed the surveyors to complete the study after Governor Eyre skilfully called for a snap vote in the House of Assembly (Stafford 2002). On one occasion during 1865 the survey activities were delayed because of the Morant Bay Rebellion that culminated in October of that year. Political conditions in the island had started deteriorating early in 1865. The island had just recovered from an outbreak of Asiatic cholera during 1853 and 1854 where some 31,000 lives were lost (Senior 1997). Economic crisis among other problems loomed. These included the unavailability of land for the small-scale planters, unemployment, low wages, irregularity in their payments and burdensome taxation. Consecutive periods of drought had ruined large quantities of provision crops and the cost of imported food such as salt fish and grain on which the peasantry depended had increased sharply because of the ongoing civil war in the USA (Black 1976). Attempts were made for both the Colonial and Imperial Governments to address the deteriorating conditions, but no attention was given. Noticing the deteriorating political conditions the surveyors as early as August began finalization of the island-scale geological map (Fig. 2.4; Sawkins and Brown 1865). The map therefore is dated 1865 but the survey reports remained unfinished.

The rebellion at Morant Bay started as a local uprising against the magistracy but soon developed into a revolt against British rule when the idea was encouraged that if the white population was exterminated the island would belong to the black peasants (Black 1976). The rebellion had resulted in many deaths and in 1866 a Royal Commission of Enquiry was established to determine, from the respected persons of society, 'the past and present disposition of the black population' (Anonymous 1866). The enquiry explored the origin and suppression of the rebellion. Senior officer in the Colonial Office, Henry Taylor had originally expressed his opposition to the Emancipation bill. When the insurrection occurred, he called the attention of Sir Charles Edward Grey, former Governor of Jamaica, explaining that it was the manifestation of what he, Taylor wanted to prevent (Anonymous 1885). A significant number of testimonies were given, among which was that from Sawkins. Sawkins provided his testimony by way of letter to the Secretary of the Executive Committee dated January 2, 1866, where he stated that conditions were noticeably deteriorating. He stated that George William Gordon, whom he had met several times, claimed that Jamaica would be dominated by blacks and there would be no whites in Jamaica by 1872. George William Gordon was an elected member to the Kingston Common Council who served as Acting Mayor on several occasions. He had convened a meeting in August with the people in Morant Bay and proposed a peaceful recourse by making a strong petition to the Parliament. Eyre attributed the rebellion to the agitation of Gordon, and so had him arrested and hanged.

Sawkins testimony presented Gordon as the instigator of the resistance against the white people in the name of religion (Sawkins 1866) which contrasts the description given at his death that he was 'a man of princely generosity and of unbounded benevolence' (Black 1976). Sawkins also narrated events from 1863 and 1865 where he was attacked by armed men while on official duties. The first incident occurred at Stony Gut, St. Thomas-in-the-East and the second at Palmetto Grove,

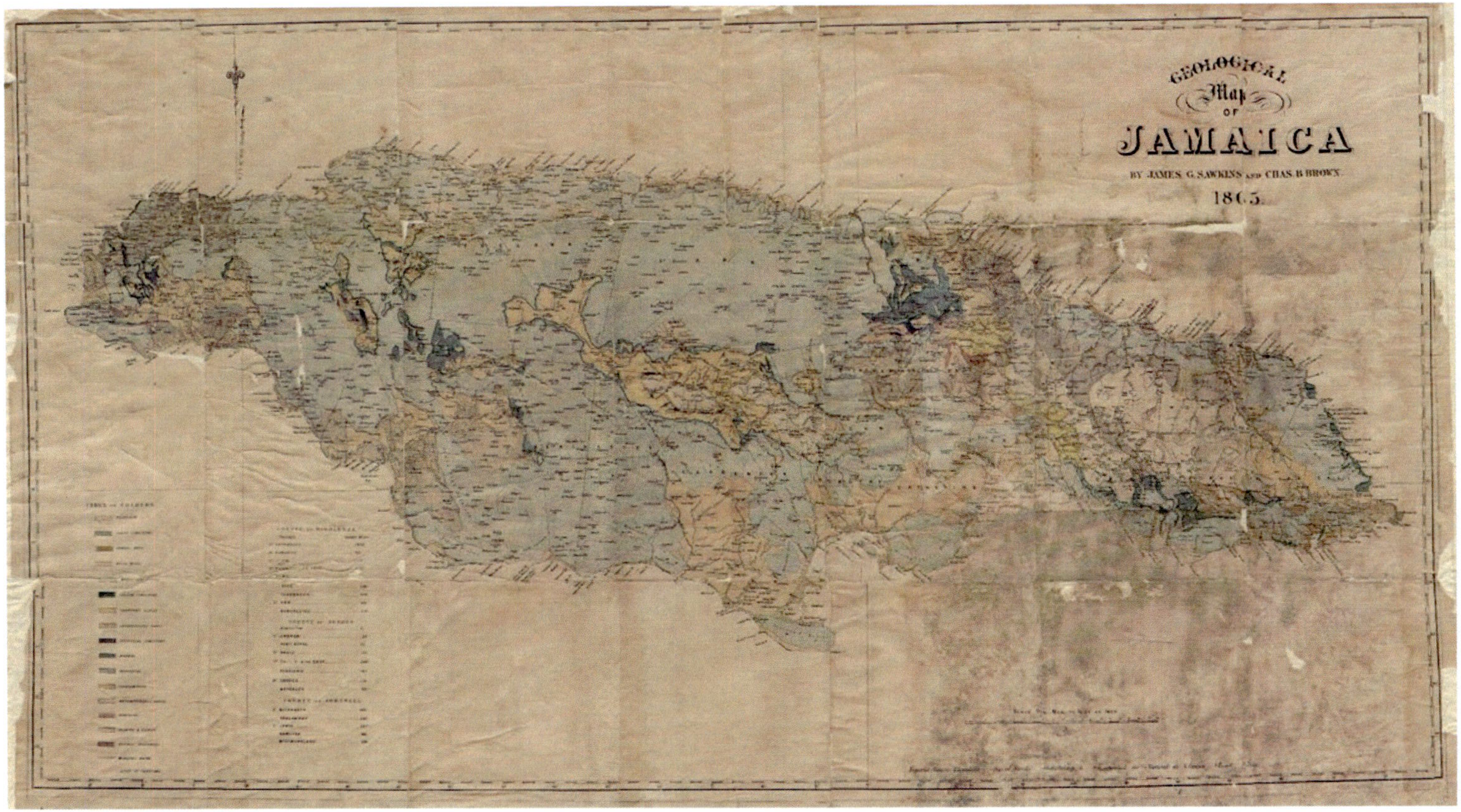

Fig. 2.4 1865 Geological Map of Jamaica prepared during the first Government-commissioned Geological Survey (Sawkins and Brown 1865; map sourced from the National Library of Jamaica, and may have been the original copy lodged with the Government of Jamaica)

St. Thomas-in-the-Vale. While in Clarendon, young men threatened to stone his party, and on the Old Harbour to Spanish Town Road he and his team were stoned. One of his servants was hurt in this incident. He mentioned that his colleague George Wall had been attacked by machete-wielding Negro peasants (Sawkins 1866). It is uncertain if the surveying party may have provoked the peasants which may occur if they entered their lands without permission or damaged their crops or property, among other things. The Commission concluded that the disturbance started as an organized resistance to lawful authority. They credited Eyre for prompt action in bringing back control but held him responsible for the prolonged use of excessive force and for how Gordon was treated and executed. He had been suspended during the investigations, and the Commission's report was detrimental to his career. He was recalled to England and dismissed from service.

It was not until 1867 before the geologists concluded the surveys of Jamaica yet the reports were only available two years later and dated as such (Sawkins 1869b). Concerned by the delays the agriculturalists in the colony in an 1868 letter to the *Gleaner* editor entitled '*Geological Survey for the Benefit of Agriculture of the Island*' queried the expected publication of the survey report (Anonymous 1868). The writer highlighted that an updating of the geological survey for Canada was ordered by the Imperial Government, yet the Jamaican reports were still being awaited. He showed that only a short time had elapsed since Canada was surveyed in the 1840s, yet they were already considering an update. The 'Reports on Geology' volume was finally published as a Memoir of the Imperial Survey (Sawkins 1869b). The 'Introductory Notice' written by Murchison served to assure all the parties concerned that the purpose of the survey was fulfilled. For the agriculturalists, the document also included an appendix entitled *Outline of the Flora of Jamaica* by Nathaniel Wilson, Scottish botanist and gardener who served as Botanist and Curator of the Botanical Gardens at Bath, St. Thomas from 1846 to 1867 (Wilson 1869 and Anonymous 2014). On February 7, 1870, through the Colonial Secretariat Law, the Jamaica Geological Survey was officially recognized as the standard geological study for the island (Aratram 1999).

2.3 Deployment After Jamaica

After mapping Jamaica, Sawkins and Brown were dispatched to the Eastern Caribbean. However, these islands could not afford to finance the activity, so the geologists proceeded to British Guiana and the survey of that colony commenced in November 1867 (Stafford 2002; Brown and Sawkins 1875). British Guiana initially had refused to accommodate the activity, but with optimism of finding gold, they withdrew their decision. In 1850 there were reports of placer gold being discovered in Venezuela in rivers that originated in Guiana, and there was also a short-lived gold rush in the region in 1857 (Stafford 2002). Just over two and a half years after the geologists were deployed, Sawkins retired in June 1870 and Brown was tasked to single-handedly conduct the surveys. Despite the challenges he endured during his expeditions, he is credited for the discovery of the Kaieteur Falls on the Potaro River. The official report on the 'Geology of British Guiana' was published in 1875 as a memoir of the Imperial Survey (Harker 1917; Stafford 2002).

Sawkins spent his retirement years back in England, residing in London until he died of a stroke in 1878 at seventy-two years. He was married twice, last to Mary Hussey Brodie who helped him prepare the illustrations of the observations from his geological surveys (Compton 2006). He was remembered as having spent nearly fifty years of his life wandering over the surface of the earth, visiting many of the continents (Sorby 1879). Charles Brown died much later in 1917 at the age of 78 years. While on his South American expeditions, after mapping Guiana, between 1873 and 1875, he travelled some 15,000 miles on the Amazon and its tributaries. After South America he worked in gemstone mining in Burma, North Carolina, Ceylon, and New South Wales. It is said that during his later years Brown suffered from the effects of the fevers he contracted while on his tropical expeditions (Harker 1917).

2.4 Calls for Update to the Survey

A little more than 20 years had passed since the findings of the first Government-commissioned geological survey were published (Sawkins and Brown 1865; Sawkins 1869b) and there were renewed calls for another, but detailed survey. The survey findings were deemed as a first start. Further, it was proposed that the geological survey activity should become a permanent establishment (Anonymous 1868; Daily Gleaner 1893). At the time, many national survey organizations of the large countries were updating their maps through systematic mapping programmes and it was deemed appropriate that Jamaica should do the same (Anonymous 1893).

With Jamaica being developed, the colonists recognized the fundamental importance of information obtained from geological surveys to guide development projects. The colonists expressed their opinion in the local newspaper, the *Daily Gleaner*. In 1891 it was proposed that water wells should be sunk in areas where there was a need, or even test wells for petroleum, and these works should be guided by the findings of the geological survey. However, the writer expressed disappointment that the scale of surveys done in the 1860s was insufficient and harshly criticized the work as '*terra incognita*' and 'little more than an undeveloped diagram outline' (Anonymous 1891). Another writer (Daily Gleaner 1893) stated that the Public Works Department had written to the Board of Directors of the Jamaica Institute (now the Institute of Jamaica) early in 1892 to query if a revision of the island's geology was planned. John D'Aeth, the District Engineer of the Department recommended an immediate geological survey considering that there was widespread construction of new roads that offer immense opportunities for observations. He stated that Manchester, Trelawny and St. Ann parishes had temporarily secured the services of a competent geologist, a Mr. Easton, who has considerable knowledge of the West India region, and he recommended this gentleman as a suitable consultant to revise the 1865 survey map. The Board of the Institute concurred with the recommendation of the engineer but stated that funds were not available to undertake the task. The matter was referred to the Government accompanied by an opinion that the proposed survey though expensive would be useful. Major General Black, administering the Government, responded in July 1893 that however valuable such a survey would be, the financial conditions of the colony at the time were unfavourable and it could not be accommodated. Black's response was described as a 'very convenient one which would be more acceptable if it were not so flippant and at the same time so stereotyped' (Daily Gleaner 1893).

The writer recognized that across the world, countries such as Britain, the USA, and Canada were updating their geological survey work. They stated that despite the work of De la Beche in the 1830s, the geological officers were engaged in further detailed examinations. In Scotland, topographic maps were corrected and new maps constructed which supported the mapping of more accurate geology. This enabled the identification of building materials, the measurement of ores and coal seams, and the location of best sources of water supply. In the USA there was evaluation of coal, mineral resources, and water supply. Canada was described as gaining the greatest benefit where vast mineral resources would have remained undiscovered had it not been the mineral exploration activities conducted by the national Geological Survey. It was expected that when the geologists completed their surveys, Canada will be found to be one of the richest mineral-bearing regions in the world. A survey for Jamaica was stated as having several economic benefits along with the discovery of ethnographical material, accurate determination of the surface features of the island, and the preparation of new topographic maps. The studies would generate useful facts for the productions of proper textbooks for the schools, which the writer stated was desperately needed. The resulting geological map would be ample return for the cost of the survey (Daily Gleaner 1893).

2.5 Independent Studies

The call for a geological survey and the establishment of a permanent department was never satisfied. Due to the scarcity of resident expertise in geology in the colony most of the research were done by foreign scientists. In 1867 a Dr. Allen had contributed geological specimens to Dr. Moritz Stern of the Royal Society of Arts and Agriculture (Anonymous 1867) to support some research. Much later in July 1892, Alfred John Jukes-Brown, stratigrapher and invertebrate palaeontologist of the Geological Museum in Jermyn Street communicated to the Institute of Jamaica a benevolent request. He wanted to know if anyone in the island would assist in an initiative towards obtaining a better understanding of the White Limestone rocks. It was his desire to ascertain the nature and mode of formation, and to test the hypothesis that the rocks are part of a double series. He had examined a few specimens and found that some were ancient coral reefs and others were deep-sea oceanic limestone. He reminded that the island's geological study had not been advanced since the first Government-commissioned survey, which '... was good for its day but very much remains to be done'. Brown requested the assistance of persons to send him specimens accompanied by notes of their position and locality. He would interpret the data collected. The Curator of Museum at the Institute, Theo Cockerell, published Brown's request in the *Gleaner* (Cockerell 1892a).

By the end of August 1892, Cockerell reported that a box of specimen was sent to Jukes-Brown, and with a Mr. Gregory of the British Museum, they analysed the rocks (Cockerell 1892b). They confirmed that some of the limestone specimens were coralline rocks which occurred at high elevations throughout the island. They further interpreted that the island was raised some 2,000 ft. during recent geological times. Not excluding the commercial potential Jukes-Brown noted that some of the white limestones were suitable as lithographic limestone. He added that lithographic limestone in England is currently imported from Bavaria and is quite expensive. By October the curator reported that several donations of specimen were received for testing (Cockerell 1892c). A Mr. Espeut sent a specimen of limestone to Jukes-Brown, but it was not suitable, whereas a 'magnificent specimen' from Buff Bay was received from a Mr. Mackinnon of the Jamaica Railway and was forwarded to England for analysis. Mr. Jukes-Brown also explained that the Bavarian supply was almost exhausted, and a new trade agreement could be organized if Jamaica had material of similar quality.

There were also notable interests in other geological disciplines. The donation of an enormous fossil rudist bivalve shell *Hippuritidae* by D'Aeth to the Institute of Jamaica in 1896 is recorded (Anonymous 1896b). The Public Works Department had unearthed the specimen during the construction of the new road from Spalding to Grantham via Logie Green in northwest Clarendon. This fossil originated from that exposure of Cretaceous rocks in the centre of the island. Similar specimens were formally described by Robert Whitfield, Curator at the American Museum of Natural History who published two papers on the rudist bivalve species in Jamaican Cretaceous rocks in the Museum's Bulletin of 1897 (Whitfield 1897a, b). These fossils were of interest at the Institute and locally, geologist Francis Nicolas presented the findings of Whitfield's research. Other popular palaeontological work at that time included that of Thomas Wayland Vaughan from the US Geological Survey who documented Jamaican fossil corals and foraminifera (Vaughan 1899). In physical geology, Canadian geomorphologist Joseph William Spencer examined the Jamaican landscape and proposed the name *Liguanea Formation* for the deposits of river sands and gravels that comprise the Liguanea or Kingston Plains which he also dated as Pleistocene (Spencer 1898). His proposed name was retained for this formation. Robert T. Hill,

American geologist of Harvard University and the US Geological Survey attempted a correction of the 1869 survey reports. He was described as the preeminent field geologist of that time and was the first American to examine in detail the geological development of the Caribbean. Hill's surveys were supported by the wealth of Alexander Agassiz, president of a Michigan-based copper mining giant who oddly enough had commissioned him to conduct studies not associated with economic geology. Hill used Sawkins' map as his base and corrected numerous errors in the geological succession. Hill was able to apply modern concepts to determine, for the first time, the geological history of Jamaica. He visited the island between 1896 and 1897 and made over 1,300 km of geological traverses (Zans 1955; Donovan 2010). He collected fossils from numerous localities, and borrowed the collection of the Jamaica Geological Survey housed in the Museum at the Institute of Jamaica for expert analyses. By early 1898 Hill reported that he had described 26 local rock formations and reviewed the fossils. He published his findings on *The Geology and Physical Geography of Jamaica* as a comprehensive memoir (Hill 1899) through the zoology museum at Harvard. He also presented a revised geological map. Hill's descriptions of the physiographic features of the island were indeed valuable, despite minimal attention to the economic geology. The scientific community appreciated this contribution which led someone to write that:

> . . . with all these contributions we look forward with assurance to the early establishment of our Geology and Palaeontology on a more extended and reliable basis than has been the case up to the present. (Anonymous 1898a)

An extract of Hill's paper appeared as the second of two papers written on Jamaica in the December 1899 volume of the Scottish Geographical Magazine. The first paper presented a general geographical description of Jamaica with remarks on some of the other West Indian islands (Nash 1899a). Accompanying this article was a coloured geological map of Jamaica. The second manuscript was the summary of Hill's work (Nash 1899b). The publication of these two articles was announced in the *Gleaner* where the writer stated that since the work of Sawkins to the time of Hill, 'nothing was done in the direction of determining the correct rock structure of the country' (Anonymous 1900).

The Kingston Earthquake of January 14, 1907, presented the opportunity for understanding the tectonic motions of the island. This quake had resulted in about 800 deaths and property loss of approximately £2,010,000 (Cundall 1927). English geographer Vaughan Cornish and his wife were (conveniently) in Kingston on the day it occurred and therefore was able to make observations and interpretations about the event which he shared at the Institute of Jamaica (Anonymous 1907). He explained that earthquakes represent 'a sudden snap' of rock strata due to creeping of the crust and that the quake was recorded on seismographs located at great distances from Kingston. The aftershocks however were recorded on the seismograph at the Public Works Department, and this was monitored by Government Meteorologist Maxwell Hall. In 1880 at the beginning of his tenure, Hall had initiated the first comprehensive study of earthquakes in Jamaica (Hall 1880).

Cornish noted Hall's observation that some of the Public Works' seismographs indicate a force coming from the position of faults that were mapped during the geological survey in the 1860s, such as that fault on the east branch of the Spanish River in Portland parish. He therefore recommended monitoring the behaviour of the land during earthquakes specifically at the geological contact of the shale formation in eastern Jamaica, namely Sawkins' Carbonaceous Shale or Hill's Richmond Formation. He further recommended that the boundaries of the formation should be transferred to a topographic map, and those in the east examined alongside the faults shown in Sawkins' geological sections. Cornish also reported recession of the sea on the north coast at Orange Bay, Buff Bay, Annotto Bay, Port Maria, and Ocho Rios where the sea subsided several feet vertically, withdrawing to a considerable distance and finally coming back in waves. He was however interested in changes in soundings off the coasts. Engineer D'Aeth told him of indications of an apparent rise of the land at the Yallahs Point salt ponds, and at Morant Bay and Manchioneal in St. Thomas parish. Detailed descriptions of the earthquake were published by Cornish and several others (Cornish 1908; Brown 1907; Hall 1907, 1909; Fuller 1907). Hall had prepared a register of all the earthquakes known in Jamaica from 1688 up to 1907 (Hall 1907). A later study by American geologist and geology professor Stephen Taber (Taber 1920) described the major earthquakes experienced in Jamaica from 1687 to 1914, and explained that the 1907 earthquake, like the 1692 earthquake that destroyed Port Royal originated on Jamaica's northern coast, and not on the south coast as most persons thought.

2.6 On Mining and Geology

Several accounts on Jamaica's potential mineral wealth were shared by various experts towards the late 1800s. This includes H. I. Burger's 1881 volume on *The Mineral Resources of Jamaica* (Burger 1881). In this work, there is mention of a piece of platinum found by an American between 1838 and 1839 in the Yallahs River. Two documented lectures were given by Rev. Horace Scotland, clergyman, amateur geologist, and friend of James Sawkins (Scotland 1890). He spoke *On the Geology of Jamaica* (Scotland 1889a) and *On Mining in Jamaica, Retrospects and Prospects* (Scotland 1889b). He too mentioned that the first known specimen of native platinum brought to Europe was found in Jamaica. This was brought to England by a Mr. A. Wood in 1741. Both these findings were never confirmed. He presented a convincing picture of the rich potential of Jamaica's mineral deposits and described the events that unfolded during the mining boom of the 1850s. Scotland reminded participants that there were no mining operations underway in Jamaica at the time. Other popular lectures included that by Frank Cundall, Librarian and Secretary of the Institute of Jamaica in 1896, who presented a paper written on *Jamaica in the past and present* (Anonymous 1896a). He gave some statistics about the island and spoke about the economic geology. He reminded the audience of the abundance of copper in the island, but that lead, zinc, manganese and gold were only to be found in small quantities. The events leading up to the failures of Jamaican mines were also discussed.

Geologist Francis Nicolas in May 1898 presented lectures on the *Economic Geology of Jamaica* at the Institute's East Street rooms (Anonymous 1898b). He lauded the work of Sawkins and the team of surveyors stating that the final reports could be read in connection with tropical American geology all over the world. Nicolas explained that Jamaica postdated the major mineralization episodes when coal measures and iron formations across the world developed. He said that the Hope Mine was only a small indication of what could have been an extensive deposit. Copper deposits he explained, might be found but only on a small scale. He warned that one should not become enthusiastic over pretty specimens of green carbonate (malachite), as it takes just 'a little copper to make a big stain, and a little bit of chalcopyrite did not by any means make a mine'. He considered the manganese prospect at Marshall's Hall, Portland to be the best deposit. However, he was informed that the land is owned by the Maroons who admitted that their women cunningly collect buckets of ore and broadcast it on the ground to make a bigger ore showing; but it was still possible to find untampered mineralization. He mentioned the abundance of good clay for bricks and pottery and in time the island can become self-sufficient with regards to these products. Lime on the other hand could be exported to South America where the demand was increasing, and a paying trade could be developed (Anonymous 1898b).

Nicolas' talk had generated an intense debate where his lecture was described as a 'dismal picture' of the mineral wealth of the island (Anonymous 1898c). Just before the lecture, a discussion ensued about the presence of gold and hopes were high. The call was again made for the establishment of a Geological Survey to settle the matter. It was stated that Nicolas had used Sawkins' research which was too general, and this had prevented him from determining exactly the metalliferous nature of the Island, especially that mining was extremely risky. Scotland had highlighted in his 1889 lectures a clever cartoon of a cross bones and skull with two captions (Scotland 1889b), one above and the other below. The first:

1855 - *Beware of Mining in Jamaica!* - 1858

And the second:

1885 - *Metals is PISUN* [poison]*!* - 1889

These periods marked the times of greatest financial lost in mining.

The numerous requests for a Government geological survey were recalled beginning with the attempts of Scotland and leading up to that of D'Aeth of the Public Works Department. The scientific community recommended that the island should have a permanent geologist, and they stated that 'we have to depend on obliging visitors like Mr. Nicolas to give us information about our own surroundings' (Anonymous 1898c). Despite Nicholas' dismal portrayal of Jamaica's metallic mineral potential, companies such as the Jamaica Consolidated Copper Company based in Boston were not deterred. The company leased with an option to purchase, over 2,000 acres of land at upper Clarendon in 1906 for exploration. The company conducted considerable prospecting for copper in 1907 at the old Charing Cross and Stamford Hill mines. Over 1,585,800 tonnes of ore were identified but the mines were not brought into production (Zans 1951).

2.7 Bad Science

The absence of an authoritative source for geological information exposed the colony to pseudo-science and quack interpretations. Two such instances are notable where individuals sought to mislead the public. The first in 1897 was water divining being conducted in Kingston to locate well sites. The diviner, a Mr. Stears was contracted by the Artesian Water Supply Company to assist their water exploration programme (Anonymous 1897a). The company was boring a well at Windsor Pen at Windward Road at the foot of Long Mountain. Stears' divining rod was a forked branch of the 'Cashaw' tree (cassia, *Acacia mimosa*). At 12–15 yards from the hole the divining rod trembled violently towards Stears' face. He explained that there was an underground stream of running water flowing at a rate of 7,000–8,000 gallons per hour. Stears was careful to explain that he could not divine water unless it was flowing. The boring had reached 62 ft. depth, but Stears recommended that it should be discontinued and the boring made at the spot he identified. He explained further that the water would be found at a closer depth than the other spot and it would rise in the pipe and overflow to form a true artesian well. Stears explained that in divining, if the rod moves upward, it is suggestive of flowing water, whereas downward movement may mean that there is an iron-bearing substance in the ground. At Windward Road, the rod had also moved downward prompting Stears to give this explanation. A member of the group at Windward Road confirmed the find and stated that there was no way that Stears could have known beforehand that something was below. But while these persons were being convinced, the practice was being dismissed in other places. In July 1897 an audit of the Urban District Council of Ampthill, Bedfordshire revealed some councillors had hired water diviner Leicester Gataker to find water. The Government geologist was consulted, and he showed that a proper supply could not be found where the diviner had indicated its existence. A long hearing followed and finally the auditors declared that the councillors should repay the £13 8s 7d expended to the diviner as there was no justification for spending public funds for employing persons exercising powers of this kind (Anonymous 1897b).

The second recorded occurrence was that of a twisted theory (and a possible scam) which involved the Hope Mine in St. Andrew. Historically, this mine was deemed one of the most productive and has the longest history of development dating back to the Spaniards. In the early 1900s, the mine was acquired by a L. B. Queensburgh Tulloch having title 'Consulting Mining Geologist and Principal of the Jamaica New Mining Bureau'. In May 1918 he reported on this mine which he referred to as the Rock Castle Mine (Tulloch 1918). He was overly confident that it could be revived. Tulloch invited 'the Interested and the Doubtful' to pay his mine a visit stating that he has 300,000 tons of low-grade ore ready for concentration, and heaps of valuable ore in two shafts. Tulloch invited persons to contact him for further information indicating that with the necessary funds and machinery the ore can be processed profitably. He pointed out that he will do further work to outline the stratigraphy and noted that his detailed assessment report is open to inspection. Five months later, a half-page 'Golden Message' by an anonymous author, probably Tulloch himself, representing the Jamaica Mining Bureau is published in the *Gleaner* with headlines:

> Important!
> Important!
> Mineral Wealth of Jamaica.
> Rock Castle Mine showing up large mineralized dyke with rich and important ore bodies
> Success Sought by Many, Secured by few. ROCK CASTLE MINE As it now Stands,
> Counting for the Millions. (Anonymous 1918)

The writer begins by boldly declaring:

> . . . the vast possibilities of the island's mineral resources, . . . are still unscratched. Rock Castle Mine now bids fair to let Jamaica feel proud of her rock formations. . .. the wind is blowing favourable now, and the wheel of fortune turns to Jamaica. A NEW ERA IN MINING will now be, and after the war is over, a great development. (Anonymous 1918)

The article boasted the determination and scientific knowledge of the mining geologist and operator Mr. Tulloch, stating that for four years he has consistently investigated the nature and character of the formation as well as the value of the ore bodies.

> Every point of the property carefully examined, and the underground boring continue of such lines to demonstrate the extent of the ore deposit, as well as the probable and positive value of the ore bodies. Numerous tests have been made on the different grades of ore and over 300 field tests and microscopic examinations have been made with the polarizing microscope, not only rich ores but also of the lowest grade and common country rock at Rock Castle. (Anonymous 1918)

Hardly, if any geological work in Jamaica up to this time, had utilized the technology of the polarizing microscope, considering that it may have been quite costly. The Bureau article consistently repeated the plea for funding, stressing the profitability of the venture. With the support of some calculations, optimism for profitability was extended to the global scene, envisioning a total change for the island by the activity of this single mine:

> With proper management and good engineering or mining knowledge point to a most hopeful near future which will not only be profitable but will place Jamaica on the map with a distinguished recognition as a financial centre of importance in the West Indies. ... ROCK CASTLE MINE IS AN INEXHAUSTIBLE CAMP AS FAR AS DEPTH IS CONCERNED. (Anonymous 1918)

The writer mentioned an iron-bearing formation with other richer and valuable ores to unreachable depth. He stated that this iron ore mined at the surface would yield profit of twenty shilling per ton on the spot. The writer cited that there was another area with thousands of tons of mixed low-grade ore with copper, iron, and gold. Additionally, he mentioned that there was over 200,000 tons of low-grade copper and iron ore at the surface and that could run five shilling per ton besides the gold by-product. This he stated was equivalent to £50,000, with processing and handling charges running at £30,000, leaving £20,000 profit. He then declared in large letters:

> ROCK CASTLE MINE HAS INEXHAUSTIBLE WEALTH. (Anonymous 1918)

In the article, the writer, never strayed too far from geology. Though he convinces his readers of the economics of the mine, he constantly revisits the geological features of the property. Interestingly enough, there was one assertion that he strongly defended: that of the age of the rock strata at Rock Castle. Citing his petrographic work on the crystalline igneous rocks found in eastern and central Jamaica he stated that the formations were as old as those in Europe and America, and strongly denounced the work of early surveyors:

> The idea and unqualified geological knowledge entertained by those who not only conjectured but asserted that Jamaica is a young formation recently rose from out of the sea are simply unacquainted or child's talk. (Anonymous 1918)

The encouragement continued with statements that Rock Castle may become a 'copper queen' in Jamaica, or even a 'gold king!' In supporting the venture the writer mentions the comments from a 'Big London firm with [an] associate [that] possesses vast knowledge in minerals':

> Rock Castle rocks wonderful, indicates immense underground possibilities, will interest themselves in the proposition. Is no lightweight talker. (Anonymous 1918)

By February 1919, Tulloch presented an article in the Gleaner directed *Only to Mining Men* (Tulloch 1919). He continued promoting the mine stating that Rock Castle Mine is an excellent opportunity and puts out a proposition. He declared that the mine:

> . . . was an old producer yet the immense underground values [had] not been scratched. Now into new hands and looming to resume as a winner – to an inexhaustible producer. (Tulloch 1919)

He sought to explain why the mine had failed in the past citing the want of proper mining knowledge, the absence of modern mining machinery and improper methods of treating medium- and low-grade ore on the spot. The absence of proper shipping facilities either to England or America, want of proper geological examinations, and ignorance of the immense ore bodies, their underground behaviour and their origin, and ore grades were also listed as reasons. Tulloch further explained that a serious obstacle in former years by old hands was reckless preliminary operations to shore up the weak rock layers. In some places the miners referred to it as 'a high-level hell'. He however, reassured his audience that special attention was devoted to 'removing all weakened and weighty-hanging rocks' and that danger spots were now safe. Visitors were now welcome. Tulloch then noted that the mine is showing signs of being a large producer and he anticipates prosperity. He then warned that if the proposition was not taken before early March, it will be taken up by London and Denver mining men for development.

It seems that the interest of the local capitalists was unmoved and whether the proposition went to London or Denver is not known. Each time an article was published there was a noticeable name change of the company. In May 1918 it was the Jamaica New Mining Bureau (Tulloch 1918), then by October it was called the Jamaica Minerals and Mines (Anonymous 1918) based at Oxford Street, Kingston. Then by February 1919 the company was called the Western Mining Bureau based at Bond Street (Tulloch 1919). After this there was no further mention of this venture in the local newspapers. The Rock Castle Mine or the Hope Mine as it is better known was reopened between 1942 and 1943 during mineral exploration activities conducted by mining engineer and Assistant Commissioner of Lands, A. H. Edwards who examined it as a potential source of lead and zinc (Zans 1951). Some years later, the British established a Geological Survey Department to conduct a major programme of geological survey and mineral exploration of the colony. The Hope Mine was among the first places to be assessed; this being in November 1949 a month after the survey was established.

References

Ancestry (2021) Lucas Barrett. https://Lucas_Barrett&birth=1837_London-Middlesex-England&death=1862_Port+Royal-Jamaica. Accessed 1 Sept 2021

Anonymous (ed) (1866) Letter—Jamaica disturbances. Papers laid before the royal commission of inquiry by Governor Eyre. In: Reports from commissioner: session Feb 1–Aug 10 1866, Twenty-two Volumes Jamaica. George Edward Eyre and William Spottiswoode, London

Anonymous (1867) Notice of presentation. Daily Gleaner, 31 Jul 1867, np
Anonymous (1868) Geological survey for the benefit of the agriculture of the island. The Gleaner, Aug 11, 1868, np
Anonymous (1885) Sir Henry Taylor, an autobiography of Henry Taylor in two volumes, with portrait. Harper and Brothers, New York, New York Times, 19 Apr 1885
Anonymous (1891) Artesian wells. Daily Gleaner, 18 Mar 1891, p 4
Anonymous (1893) Geological work in other countries. Daily Gleaner, 8 Sept 1893, p 4
Anonymous (1896a) Mr Cundall on Jamaica. Daily Gleaner, 11 Jan 1896, p 4
Anonymous (1896b) A giant fossil. Daily Gleaner, 24 Oct 1896, p 4
Anonymous (1897a) Water divining in Kingston. Tri-Weekly Gleaner, 1 Apr 1897, np
Anonymous (1897b) Costly faith in water divining. Daily Gleaner, 25 Jun 1897, np
Anonymous (1898a) Jamaican Palaeontology. Daily Gleaner, 29 Jan 1898, p 4
Anonymous (1898b) Lecture at the Jamaica Institute. Daily Gleaner, 6 May 1898, np
Anonymous (1898c) Gold in Jamaica. Daily Gleaner, 7 May 1898, p 4
Anonymous (1900) The geological history of Jamaica. Daily Gleaner, 14 Jan 1900, p 4
Anonymous (1907) The Kingston earthquake in Jamaica: its causes and effects. Daily Gleaner, 21 Jun 1907, p 9
Anonymous (1912) Obituary of George Parkes Wall. Q J Geol Soc 68:lxiii–lxiv
Anonymous (1918) Important! Mineral wealth of Jamaica. Daily Gleaner, 26 Oct 1918, p 17
Anonymous (2014) Fern Gully, Jamaica. http://ferngullyjamaica.synthasite.com/n-wilson.php. Accessed 4 Mar 2014
Aratram MR (1999) 140 years of geological assessment in Jamaica—perspectives from the Mines and Geology Division (MGD). Daily Gleaner, 9 Aug 1999, p B4–B5
Barrett L (1860) On some Cretaceous rocks in the south-eastern portion of Jamaica. Q J Geol Soc 16:324–326
Barrett L (1869) Report of Lucan Barrett, Director of the West Indian Geological Survey—1862. In: Sawkins JG (ed) Reports on the geology of Jamaica, or Part II of the West Indian Survey, with contributions from G. P. Wall, Lucas Barrett, Arthur Lennox and C. B. Brown, and an appendix by R. Etheridge. (Memoirs of the Geological Survey), Longmans, Green, and Co, London, p 87–88
Barrett L and Sawkins JG (1869) Official correspondence and geological report on part of the Parish of St. Thomas-in-the-East and Portland—1859. In: Sawkins JG (ed) Reports on the geology of Jamaica, or Part II of the West Indian Survey, with contributions from G. P. Wall, Lucas Barrett, Arthur Lennox and C. B. Brown, and an appendix by R. Etheridge. (Memoirs of the Geological Survey), Longmans, Green, and Co, London, p 72–87
Black CV (1976) History of Jamaica. Collins Clear-Type Press, Great Britain
Brown CW (1907) The Jamaica earthquake. Pop Sci Monthly 70:385–403
Brown CB and Sawkins JG (1875) Reports on the physical, descriptive and economic geology of British Guiana. Her Majesty's Stationery Office, London
Burger HJ (1881) The mineral resources of Jamaica. In: Sinclair AC (ed) The Handbook of Jamaica for 1881. Government Printing Establishment, Kingston, p 55–60
Chubb LJ (1964) Lucas Barrett, a biography. J Geol Soc Jam (Geonotes) 6:3–45
Cockerell TDA (1892a) Notes from the Museum No. 16. Daily Gleaner, 4 Jul 1892, p 3
Cockerell TDA (1892b) Notes from the Museum No. 22. Daily Gleaner, 22 Aug 1892, np
Cockerell TDA (1892c) No title. Daily Gleaner, 7 Oct 1892, p 2
Compton B (2006) James Gay Sawkins. http://freepages.rootsweb.com/~dbcompton/genealogy/sawkins/i0001346.htm. Accessed 29 Aug 2021
Cornish V (1908) The Jamaica earthquake (1907). Geog J 31(3):245–276
Cundall F (1927) Handbook of Jamaica for 1927. Government Printing Office, Kingston, pp 30–45
Daily Gleaner (1893) Revision of the island geological survey. Daily Gleaner, 24 Aug 1893, p 4
Donovan SK (2010) R.T. Hill (1858–1941) and The geology and physical geography of Jamaica: Study of a type of Antillean development (1899). Mem Geol Soc Am 205:47–57
Fuller ML (1907) Notes on the Jamaica earthquake. J Geol 15:696–721
Hall M (1880) Notes on hurricanes, earthquakes and other physical occurrences in Jamaica, up to the commencement of the Meteorological observations. Jamaica, Meteorological observations 2, Government Printing Establishment (1896)
Hall M (1907) Third report on earthquakes in Jamaica—the great earthquake of January 14th 1907, and the After-shocks. Weather Report No. 337, Jamaica, Meteorological observations 4, Government Printing Office (1911)
Hall M (1909) Further notes on the great earthquake, 1907, Jan. 14. Weather Report No. 365, Jamaica, Meteorological Observations 4, Government Printing Office (1911)
Harker A (1917) Obituary of Charles Barrington Brown. Q J Geol Soc 73:lxiv–lxv
Hill RT (1899) The geology and physical geography of Jamaica: study of a type of Antillean development. Based upon surveys made for Alexander Agassiz. With an appendix on some Cretaceous and Eocene corals from Jamaica by T. Wayland Vaughan. Bull Mus Comp Zoo Harvard College 34

Jamaica Gleaner (2002) Pieces of the past, Jamaican rum—a kill-devil of a drink. The Gleaner Company Ltd., 16 July 2002. http://jamaica-gleaner.com/pages/history/story0030.html. Accessed 9 May 2010

Lowe C (1892) Four national exhibitions in London and their organiser. With portraits and illustrations. T. F. Unwin, London

Mitchell S (2016) The geologists who studied Jamaica—*Lucas Barrett, F.G.S., F.L.S.* https://www.sfmgeology.com/jamaicangeologists/barrett.html. Accessed 1 Sept 2021

Nash AG (1899a) Jamaica, with remarks on some of the other West Indian Islands. Scott Geogr Mag 15(12):617–627

Nash AG (1899b) Sketch of the geology of Jamaica. Scot Geog Mag 15(12):628–639

Prestwidge W (2008) Jamaican Parish reference. http://web.archive.org/web/20110515000000*/, http://prestwidge.com/river/jamaicanparishes.html. Accessed 6 Dec 2021 (Site is now archived)

Ramsay AC (1864) The anniversary address of the president. Q J Geol Soc 20:xxxiii–xxxiv

Sawkins JG (1866) Letter—Jamaica Disturbances. Papers laid before the Royal Commission of Inquiry by Governor Eyre. In: Anonymous (ed) Reports from Commissioner: Session Feb 1–Aug 10 1866, Twenty-two Volumes Jamaica. George Edward Eyre and William Spottiswoode, London, p 118–119

Sawkins JG (1869a) Geological report of the island of Anguilla—1865. In: Sawkins JG (ed) Reports on the geology of Jamaica, or Part II of the West Indian Survey, with contributions from G. P. Wall, Lucas Barrett, Arthur Lennox and C. B. Brown, and an appendix by R. Etheridge. (Memoirs of the Geological Survey), Longmans, Green, and Co, London, p 257–262

Sawkins JG (1869b) Reports on the geology of Jamaica, or Part II of the West Indian Survey, with contributions from G. P. Wall, Lucas Barrett, Arthur Lennox and C. B. Brown, and an appendix by R. Etheridge. (Memoirs of the Geological Survey), Longmans, Green, and Co, London

Sawkins JG, Brown CB (1865) Geological map of Jamaica—scale: one mile to ¼ of an inch. Vincent Brook, Day & Son Lith., London

Scotland H (1889a) On the geology of Jamaica. In: Lectures delivered under the auspices of the Institute of Jamaica, 1882–1889, Kingston

Scotland H (1889b) Mining in Jamaica, retrospects and prospects. In: Lectures delivered under the auspices of the Institute of Jamaica, 1882–1889, Kingston

Scotland H (1890) Institute of Jamaica, Popular lectures, fifth series, lectures first and third, first: on the geology of Jamaica, given at the Institute Rooms, Kingston, on Wednesday 6th February, 1889, third: on mining in Jamaica, retrospects and prospects, given at the Mico Institute, Kingston, on 5th March, 1889. Mortimer C. DeSouza, Jamaica

Senior CH (1997) Asiatic Cholera in Jamaica (1850–1855). Jam J 26(2):25–42

Spencer JWW (1898) Late formations and great changes of level in Jamaica. Trans Can Inst 5:324–357

Sorby HC (1879) Obituary of James Gay Sawkins. Q J Geol Soc 35:54

Stafford RA (2002) Scientist of empire: Sir Roderick Murchison, scientific exploration and Victorian imperialism. Cambridge University Press, London

Taber S (1920) Jamaica earthquakes and the Bartlett Trough. Bull Seismol Soc Am 10(2):55–89

Tulloch LBQ (1918) Jamaica mineral bearing formation, ore deposits of the island. Daily Gleaner, 29 May 1918, p 16

Tulloch LBQ (1919) Only to mining men. Daily Gleaner, 8 Feb 1919, p 18

Vaughan TW (1899) Some Cretaceous and Eocene corals from Jamaica. Bull Mus Comp Zoo Harvard College 34:227–250, 255–256

Wall GP, Sawkins JG (1860) Report on the geology of Trinidad; or Part I. of the West Indian Survey. Longmans, Green, and Co., London

Whitfield RP (1897a) Descriptions of species of Rudistae from the Cretaceous rocks of Jamaica West Indies, collected and presented by Mr F. C. Nicolas. Bull Am Mus Nat Hist 9:185–196

Whitfield RP (1897b) Observations on the genus Barrettia Woodward, with descriptions of two new species. Bull Am Mus Nat Hist 9:233–246

Wilson N (1869) Outline of the Flora of Jamaica—1867. In: Sawkins JG (ed) Reports on the geology of Jamaica, or Part II of the West Indian Survey, with contributions from G. P. Wall, Lucas Barrett, Arthur Lennox and C. B. Brown, and an appendix by R. Etheridge. (Memoirs of the Geological Survey), Longmans, Green, and Co, London, p 263–291

Wilson HE (1985) Down to Earth: one hundred and fifty years of British Geological Survey. Scottish Academic Press Ltd., Edinburgh

Zans VA (1951) Economic geology and mineral resources of Jamaica. Department Bulletin 1, Geological Survey Department, Kingston

Zans VA (1955) Geology and mining in Jamaica. Daily Gleaner, 10 May 1955, p 8

3 A Temporary Survey

Abstract

Facing the crisis of a water supply shortage due to drought the Jamaican Government turned to exploring for groundwater. The Government established a Geological Survey to conduct exploration of the subsurface resource. Unfortunately, it was over 50 years since it had last commissioned an entity for this research activity. As such, there was minimal up-to-date and practical authoritative geological information to begin with, and the geologists deemed it necessary to conduct broad-based surveys to understand the characteristics of the resource they were called to examine. The administrators were oftentimes dissatisfied with the rate of progress and repeatedly reduced the fiscal allotments to support the work. After four years of operation the Survey was terminated. The chief geologist, however, recognized a knowledge gap and so visited Jamaica three times thereafter for the next twenty years to conduct personal surveys. He ultimately wanted to prepare, for the benefit of the colony, a modern manual on the geology of Jamaica, and a detailed geological report on the region around the capital Kingston. Only the latter was published, but not within his lifetime.

3.1 Contemplating a Survey

As early as the 1880s it was common for British colonial administrators to frequently hire geologists, particularly to serve in their Mines or Public Works Departments, such as in Guyana in 1889. Some of these evolved into independent Geological Surveys. These included Malaysia in 1912, Ghana in 1913, Uganda in 1918, Nigeria in 1919, Malawi in 1921, Tanzania in 1925, Sierra Leone in 1926, and Kenya in 1933. Recruitment was undertaken by the Colonial Office in London, though they were employed as public servants of the respective colonies. Staff movement and promotion were frequent as vacancies in other surveys arose, and the geologists would rarely spend their entire career in one colony (Wilson 1985). Despite this global momentum, nearly 60 years had passed since the first Government-commissioned geological survey activity was conducted in Jamaica (Sawkins 1869), and still there was no efforts towards another systematic survey. Some studies, however, were done through private funding. These were specific assessments and oftentimes lacked practical application to the development of the resources of the colony.

The water resources of Jamaica were always of major concern considering that the agriculture

S. Bhalai, *The History of Mining and Geological Surveys in Jamaica*, Historical Geography and Geosciences,
https://doi.org/10.1007/978-3-031-42604-9_3

sector and the population centres depended on having a reliable supply. In 1914 the island experienced severe drought, as only 22 cm of rain were recorded at the metereological station at the Public Works office in Kingston. This prompted the need to investigate the underground resources (Matley et al. 1951). The Island Government acknowledged that a survey to investigate water resources was desperately needed and in 1921 a vote on *Geological Surveys for Water* was tabled (Anonymous 1921a). Professional Advisor to the Colonial Office in Geological Matters, Dr J. W. Evans recommended that a resurvey of the geology should be undertaken 'as soon as may be practicable' through the employment of a 'first-rate Geologist, preferably one who has intimate knowledge of the local geology' (Anonymous 1921b). The administrators lacked understanding of what a proper geological survey involved, and Evans it seems, did not provide enough details. It seems as if the general understanding was that the geologist would simply 'look for water', without ascertaining the geological conditions that impact the resource and its reliability and sustainability. It had been over 50 years since the last Government-commissioned geological survey (Sawkins 1869).

The vote for a *Geological Survey for Water* was approved and Dr Charles Alfred Matley, a 55-year-old geologist landed on October 18, 1921 (Fig. 3.1) (Anonymous 1922a). This was his first time in Jamaica. Matley was born, educated, and started his career in Birmingham. He joined the Civil Service in 1884, first working in the General Post Office, and then the War Department in Birmingham. While living there he had married Annie Loach and started studying geology through evening classes at Midland Institute. He obtained his Bachelor of Science degree in geology with First Class Honours when he placed first in the London University examinations of 1894. He later obtained his D.Sc. in 1902 and lectured geology at Birmingham and at Midland Institute. Matley returned to the War Office and in 1913 was transferred to India as Auditor of the Indian Ordnance Factories. During that time, he made discoveries of dinosaur fossils on a factory premises in Jabalpur. Throughout his career he was stationed in Birmingham, Dublin, Edinburgh, London, and then India. Matley retired in 1920, and with his wife, he left India after living there for seven years. The couple visited South Africa and England before landing in Jamaica in October 1921 where Matley had

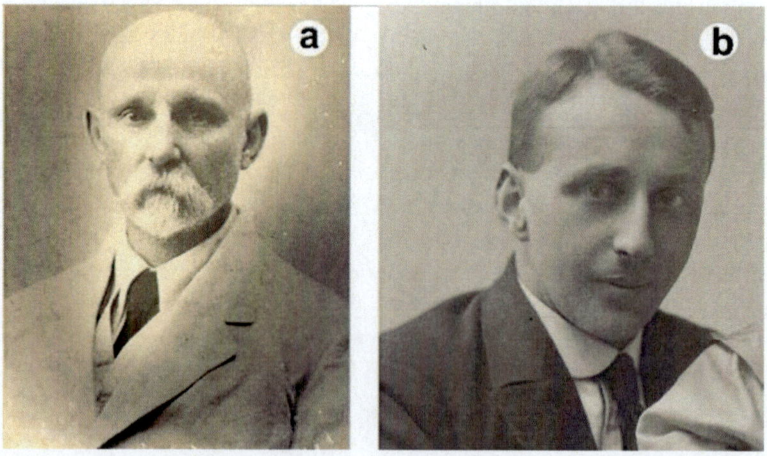

Fig. 3.1 Geologists of the second Government-commissioned Geological Survey in the 1920s. **a**—Charles Alfred Matley (Courtesy of the Mines and Geology Division, Jamaica) and **b**—Gordon Murray Stockley (Courtesy of Mr. Grenville Williams and Mrs. Dorothy Williams [nee Stockley; niece of Gordon])

accepted employment as Government Geologist (Garwood 1948; Robinson 1996). The Governor, His Excellency Sir Leslie Probyn, employed R. F. Perkins as a Geological Assistant for Matley as stipulated by the Colonial Office, and by November, Matley was deployed on the Rio Minho Irrigation System project of the Public Works Department to provide technical advice. Perkins was deployed in Trelawny to undertake water investigations in that parish. To support the investigations the English Government equipped Matley with drilling apparatus capable of penetrating 600 ft. and a foreman borer. The driller and rig arrived in February 1922 (Anonymous 1923a).

3.1.1 Deliberating the Cost

In March 1921, the Geological Survey vote was tabled for £3500 (Anonymous 1921a). However, the deliberations that unfolded over the next three years suggest that this approval was given with reluctance. The Legislature was desperate to develop the water resources of the colony and felt compelled to conduct the activity. During the initial deliberations of 1921, the administrators had questioned the colony's ability to afford a geological survey programme, but approval for *Geological Surveys of the Island for Water* was given on May 12, 1921 (Anonymous 1921c). By May the next year the tabled vote was significantly reduced to £2800, with a proposal to reduce even further to £1,800 on the basis that such large expenditure could not be afforded by the colony (Anonymous 1922a). A member of the Legislature had met with Dr. Matley who had managed to successfully convince him of the necessity of geological surveys. The geologist explained that along with the water supplies, the other natural resources should also be investigated. Matley cautioned though that Jamaicans, including the administrators, should not expect to see immediately the results of their spending (Anonymous 1922a).

In the debate of May 1922, the Colonial Secretary Hon. Brevet Colonel Herbert Bryan presented an elaborate argument for the survey (Anonymous 1922a). He cited that a cut in funding would not permit having the service of the geologist and the foreman borer and workmen assigned to the drill rig. The assistant, Mr. Perkins, had already been relieved of duties in March to cut costs. Colonel Bryan stated that the reduction would also be a waste of the money already spent. He justified his statements citing successes made during the surveys, among which were the location of good supplies of water at Montego Bay, and high-quality road material in St James and St Thomas parishes. He also mentioned that bands of what appeared to be coal were discovered in Portland, and the geologist was requested to examine them but found that they were seams of lignite. A member followed up by asking if there were any hopes of finding coal to which the Colonial Secretary smartly stated, 'not yet'. Another member supported the Colonial Secretary by confirming that since the employment of Dr Matley 'he has done very useful work and it would be a mistake to arrest the progress of the work' (Anonymous 1922a). He added that the survey had made borings for water in St. James and Trelawny, and Matley was conducting work on a major dam and reservoir construction project called the Hermitage Scheme, one of the major water supply proposals that was being considered in the country. The encouragements given by both the Colonial Secretary and the elected member were favourable to the other politicians and the members voted for the £2800 with 15 *Ayes* and one *No* (Anonymous 1922a).

The Hermitage Scheme was planned as a major project to provide an adequate and reliable supply of water through a dam and impounding reservoir. On July 19, 1922, the Legislature contemplated raising £189,000 to finance the project (Anonymous 1922b). Their vote of approval was however deferred until the report of the Government Geologist was submitted to Governor Probyn who had assigned Matley to this investigation early in 1922. The assessment took several weeks and included hydraulic tests for permeability in the boreholes. Matley reported that the site was quite suitable, and in every way, he confirmed the original opinions of the engineers (Anonymous 1923b).

By April 1923 it seems as though the work programme of the geologist had broadened, as the budget line item was now labelled *Geological Survey for Economic Purposes* with a tabled sum of £2,100 (Anonymous 1923c). The administrators, however, were still reluctant to sustain the programme and some members of the Legislature questioned when the Survey would actually end. The dissatisfaction it seems, may have been more towards Matley's performance. The Colonial Secretary explained that there were plans to search for additional supplies of water, but this would take time. When the water resources were discovered, there seems to have been little to no efforts towards further development and production, as the members recognized that no technical advice was being sought in lifting the water to the surface. This non-progressive action resulted in the members questioning the achievements of the geologist, but unfortunately no details could be provided. It seems the value of having Matley versus another geologist was being debated and so it was suggested that a progress report should be obtained from him to show the achievements up to the time, and it could be published for all to see. The members approved the vote with 13 *for* the Survey and two *against* its continuation. Consideration was also being given to hiring another geologist.

In keeping with the suggestion of the members of the Legislative Council, the Colonial Secretary sent a request dated May 7, 1923, to Matley for a progress report of the geological survey activities in Jamaica. The report (Matley 1923a) was submitted, and a copy was sent to the *Gleaner* where it was published on July 5, 1923 (Anonymous 1923a). In his report Matley was careful to document that he had been employed for one and a half years and not three years as the newspaper had erroneously reported. This was very important considering that the work he had accomplished in the time in Jamaica was under enquiry by the administrators. He mentioned the staffing assistance given to him by the Governor and the status of their employment. Perkins was relieved in March 1922 and the foreman borer, who at the time of the report, was back in England. The Public Works Department was now managing the Survey's drill rig since April 1922. Matley maintained that he was still responsible for selecting the sites for examination and determining the depths of the holes. It was Matley who had proposed the staff cuts and transfer of the equipment to curtail costs. He decided to work alone 'in view of the depressed state of the colony's finances at the time having been accepted by the Government' (Anonymous 1923a).

Matley clearly indicated that his work programme was fixed by Governor Probyn. He listed the limitations he faced which included the outdated 1865 geological map (Sawkins and Brown 1865). He states that it 'is an excellent production for its time, and equal to the standard of many of the Geological Survey maps of the British Isles of the same period' (Anonymous 1923a). The mapping work of Harvard professor, Robert Hill was also cited, but Matley notes that the detail on this map is less than the 1865 map. He also questioned Hill's corrections to geological structure and challenged his ideas on the distribution of groundwater resources. However, he positively credited Hill's corrections to the regional geology of eastern Jamaica and noted that his memoir was of great value to him. Matley also stated that the progress of his work was greatly hampered by the availability of suitable topographic maps (Anonymous 1923a). He stated confidently that if good topographic maps were available, the work in some parishes would have taken half the time and done much better. Only unpublished one inch to the mile parish maps and cadastral maps were available except for Kingston. He noted that these maps were lacking in topographic data having few natural features but no contours or hill-shading. The Kingston area had been surveyed by the War Office between 1909 and 1912, and four maps were published in 1915. The maps were dubbed the 'Pomeroy maps' after Capt. E. J. Pomeroy of the West India Regiment who supervised the topographic surveys. This military map was scaled at two and a half inches to the mile and had contours at 100-foot intervals (Anonymous 1923a). It was of great usefulness to Matley in his work in this region and formed the basis for his geological interpretations presented later (in Matley et al. 1951). He mentioned that advanced topographic surveys were underway in Trinidad and the Republic of Santo Domingo (Dominican Republic) (Anonymous 1923a).

Matley provided another report on the progress of his survey for the Autumn Session of the Legislative Council, which was also published in the *Gleaner* of November 21, 1923 (Anonymous 1923d). This may have been through another request of the Legislative Council considering the short timespan between reports. At the April 1924 sitting of the Legislative Council £1480 was tabled as the proposed expenditure for Geological Surveys (Anonymous 1924a). The discussions revisited the benefits to the island of having a Geological Survey. The Colonial Secretary, referring to the recent report, indicated that it would be a waste of money to break the progress of work by terminating the Survey. He added that the Survey had progressively identified valuable sources of water supplies, and potential traces of petroleum have been discovered in St. Ann. Not moved by the information Hon. Joseph Phillipps Henriques, elected member of St. Thomas who queried the benefit, said, '... this thing has been going on for years' (Anonymous 1924a), and he further called for the support for it to end.

The Governor, the President of the Council, gave support to the work, pointing out that if the Survey was stopped it would cost a great deal to restart it. Henriques continued to explain his dissatisfaction, wisely negating the reasons put forward by the Governor. Another member Hon. Rev. W. T. Graham representing St. Mary joined the debate slyly noting that he had personally spoken to Matley about the benefits of the Survey, but the geologist had been elusive in his reply. Graham stated that the island was not benefiting and asked for a replacement officer from another department. On this point the Colonial Secretary indicated that the new Assistant Geologist Gordon Murray Stockley is highly qualified to carry out the rest of the surveys (Fig. 3.1). Stockley had landed in Kingston on September 10, 1923. He was a qualified chemist, as well as a geologist who attended and worked at the Royal School of Mines. To restrict the debate, the Governor suggested deferring the tabled Item until the Colonial Secretary was able to comment on the present state of the finances of the colony, as at that time Jamaica was facing declining revenues (Anonymous 1924a).

At the May 1924 sitting of the Legislative Council the members were informed that Matley was sent on administrative leave, the end of which would coincide with his termination (Anonymous 1924b). Stockley the assistant was appointed head on April 18. Hon. D. T. Wint representing St. Ann moved to significantly reduce the vote to £480. However, the Secretary reported that the assistant was worth keeping since he was working on two important water schemes in dry districts of Clarendon and St. Elizabeth. He was also useful in identifying sources of suitable stone for the construction and maintenance of the island's road network. Immediately, the members protested that the road portfolio was the duty of the Public Works Department, however, the Secretary reminded that the department did not have a competent officer who was able to identify specific areas for quarrying. He added that the work of the geologist saves the country far more than what his salary amounted noting that after the water supply investigations the geologist would investigate the origin of the 'Yallahs smell'—a stench that was affecting the residents of Kingston. He mentioned that Stockley was preparing a geological map of the Kingston district and would be investigating potential deposits of chromite in northwest St. Thomas. He however pointed out that the services of the Assistant could be ended with three-month notice or by paying £100, the equivalent of three-month pay. His passage back to England was also the responsibility of the Government. Matley was already on his one-month leave and at the end there will be a charge against the vote of £1,480, if approved, for two months' salary (Anonymous 1924c).

Despite the outstanding tasks, there was continued pressure for a removal of the vote. Hon. P. W. Sangster representing St. Elizabeth expressed concerns that the colony was spending on continuing surveys which were not important to the country at the present time. He pessimistically noted that 'from an economic standpoint there was very little chance of anything being found for the benefit of Jamaica' (Anonymous 1924c). Wint again pressed for a reduction of the vote, this time by £480 though he was hoping for support from the other members for his original motion to reduce it by £1,000, on the grounds that he did not think they should embark on heavy expenditure for surveys when certain activities would not be done. He stated this because St. Ann had entered a scheme of its own with assistance from the Government, to install tanks for supplying water. There was division on the vote of £1480; however, the members decided through one *Aye* and ten *Nays* that the original figure should stand. The proposed amendment of Wint for the £1,000 allotment was thereby accepted.

3.1.2 The Work Programme

The large and growing population of the capital Kingston demanded an adequate supply of water, and several options were being considered. This included the Hermitage dam and reservoir, a major earth dam project above Constant Spring in the Wag Water River Valley. In 1922 the Kingston General Commissioners requested of the Governor for the geologist to give his professional opinion on the prospective site (Anonymous 1922b). He conducted boring tests of the permeability of the rock in the valley and found them satisfactory, and therefore gave his approval. He however, noted that the steep valley slopes were susceptible to erosion and therefore would require precautions against heavy silting. Following up with the work in Kingston and St. Andrew, the Governor in early 1923 also directed Matley to study the hydrology of the entire Liguanea Plains to irrigate it. Deputy Director of Public Works Neville Roots had suggested to Matley two proposals of increasing water supply by pipeline tunnelling into adjacent watersheds. The idea of diverting water from the Yallahs River through a tunnel to the Hope River basin was deemed by Matley as having more value. Water would be transmitted from the upper part of the Green River tributary of the Yallahs River Basin to supply a minimum flow of 10,000,000 to 12,000,000 gallons per day with potential for energy generation of 2000–3000 continuous horse-powers, all developed at comparatively low cost (Matley 1923b).

On the rock types in the island, Matley, in his first report to the Government indicated that Jamaican limestone can be separated geologically into three groups: a lower 'Yellow Limestone', a middle 'White Limestone' all overlain by another series of limestone and marls divisible into several formations. These he collectively referred to as 'Coastal Limestone' which includes Sawkins' 'White Marl' formation. He declared the White Limestone as the most important because it was over 2,000 ft. thick and covers more than half of the island (Anonymous 1923a). He also noted that the White Limestone 'areas are superficially waterless, yet are magnificent gathering grounds for water, great quantities of which are concealed below the surface' (Anonymous 1923a). Matley discovered that the normal principles governing the distribution of groundwater in other formations do not apply as there is no regular water table or levels of saturation. The water collects in subterranean channels leaving intervening areas comparatively dry. He considered the Yellow Limestone equally important because bands of clay restrict the flow of water and induce the emergence of surface springs. He maintained that the economic depth for extracting water supplies for small villages is 300 ft.; the same depth used in the days of slavery.

Matley attempted to subdivide the White Limestone into zones to ascertain geological structure and depth to the underlying Yellow Limestone that retained the water. He collected fossils to define smaller zones but noted that despite the great advances in Palaeontology in recent years, no research facilities existed in Jamaica to undertake analyses. He found that a group of fossils known as 'large foraminifera' were the best to use for the purpose of studying the White Limestone. He found that many of the genera and species have limited geological time range, and foraminifera were already being applied as zonal indicators in other parts of the world (Anonymous 1923a). As such he took his specimens to Washington to compare them with a unique collection of Tertiary fossils from all over the Western hemisphere housed in the United States National Museum. At the same time a group of palaeontologists were rigorously engaged in zoning the Tertiary formations of the southeast USA, the West Indies, and Central America. Among them were Thomas Wayland Vaughan and Wendell P. Woodring, scientists who gave Matley much needed assistance and later became his friends (Anonymous 1923a). Vaughan once worked as field assistant to Robert T. Hill who had conducted a notable survey of Jamaican geology (Hill 1899). Vaughan however had never visited Jamaica but had travelled to many of the other Caribbean territories and showed great interest in resolving the geological make-up of the region. At the time of Matley's visit, Vaughan was Director of the Scripps Institution of Oceanography at La Jolla, California, and was a leading expert in the larger foraminferal group of fossils. He would later describe a new species of foraminifera, *Discocyclina perkinsi* found in Trelawny, which was named after Matley's assistant, Perkins who conducted drilling investigations in that parish (Robinson 1996). Woodring was also a recognized palaeontologist who would later produce standard references on Jamaican fossils (Woodring 1925, 1928). Matley spent a month in the USA working on his fossils and resolving other Jamaican geological problems (Anonymous 1923a).

Both Matley and Stockley evaluated deposits of fossil fuels, minerals, and therapeutic waters all across the island (Anonymous 1922a; Matley 1924a). Lignite seams in Portland discovered by a Mr Eveleigh, and at Farlands near Aberdeen, on Crown Lands of the Cockpit Country of St. Elizabeth were both examined. The Portland seams were deemed uneconomic. The seam at St. Elizabeth was the thickest known in the island at nearly two metres. This was discovered in the Yellow Limestone in 1923 by a Mr Hewson, Chief Engineer of Jamaica Government Railway. Matley inspected the seam in September of that year collecting specimens for analysis which he submitted to the Deputy Island Chemist. The chemist found that there was too much sulphur for it to be used as fuel, and it had too little nitrogen for use in the manufacture of ammonium sulphate. The geologist calculated that only a third of the seam was workable, with the remainder being clayey lignite and lignite-rich clay, too impure for use as fuel. His observations led him to conclude that sporadic distribution of lignite beds across the island is expected as these deposits are not formed in the same way as the larger coal and lignite beds of the world. The latter is formed in swamps where carbonaceous matter accumulates. However, Jamaican lignite deposits are marine deposits of accumulated driftwood and plant material swept out to sea during floods. This situation he noted was occurring at the mouth of the Black River.

Matley documented his findings on mineral springs in his survey reports (Matley 1924b; Anonymous 1924d). At that time, an outdated and out-of-print pamphlet by J. C. Phillippo on *The mineral springs of Jamaica* published by the Institute of Jamaica (Phillippo 1881) was the most recent work. On the Island's petroleum potential Matley recalled the work of F. C. Nicolas who had said that the northwest corner of the island had potential for oil. However, two petroleum geologists had examined the area and reported that no further exploration should be conducted. Matley concurred with their conclusion noting that the area is likely to be unfavourable. He however discovered a methane seep of about 600 L per day at Brine Spring at Windsor near St. Ann's Bay. Here he recommended testing for oil citing the presence of salt water and methane gas as the usual accompaniments to petroleum. He noted, however, that there was no oil observed in the seepage of gas and water. On the other hand, he stated that even if the presence of petroleum is highlighted by boring, Jamaica is unlikely to ever become an oil-producing nation considering the general absence of surface indications. He cited Cuba where there are oil seeps, asphalt and gas seeps reported in every province but there is no production from any wells.

Matley was hopeful that his colleague Gordon Stockley, who specialized in mineralogy and ore petrology would get a chance to examine specific rock types having economic value before the Survey was discontinued. Stockley examined, an outcrop of serpentinite, a dark green metamorphic rock, located at Arntully, which had been known since the 1860s Survey. He had found through microscopy, crystals of chromite and a chromium-bearing spinel which was confirmed by chemical analyses. He expected that this body of rock would be the Jamaican equivalent to the Cuban serpentinite where there are extensive deposits of workable chromite (Stockley 1925b). Stockley pointed out that the Blue Mountain countryside had never been seriously examined for minerals though legend has it that the Spaniards were known to have carrying parties of pack mules laden with gold coming from Cuna Cuna Ridge towards Port Royal. Apart from the metallic minerals, the geologists provided support to the Public Works Department in identifying suitable areas for quarry stone. Matley was certain to mention the dykes and intrusive sheets of basalt and fine-grained dolerite at Johns Hall district, St. James and in Hanover, south of Lucea which he expressed should produce first-class road stone (Matley 1924b).

3.1.3 The Kingston District

The survey of Kingston in 1923 was the most thorough investigation by the geologists, and the report on the hydrology of this district was published in the *Jamaica Gazette* of August 10, 1923 (Matley 1923a). The geologists cited the availability of an accurate topographic map, the 'Pomeroy' map, as the main factor that facilitated a proper geological survey. The surveys included water studies for the Hermitage and Yallahs projects, road stone assessments, and the evaluation of a section of the Palisadoes for a proposed waterway into Kingston Harbour on behalf of the Admiralty. The mapping stage for Kingston was only partly complete up to the time of Matley's dismissal. He agreed with Stockley to cooperate on the production of the final map and the explanatory memoir. The geologists noted that

> Most of the formations of Jamaica and some metalliferous areas are present [or may be observed] in the Kingston district. Its geology is almost an epitome of the geology of the whole island, and the map and memoir should be of scientific and practical interest to Jamaica and visitors. (Matley 1924b:14)

Matley would continue to refine his geological interpretation of Kingston up to the time of his death in 1947 (Robinson 1996).

3.1.4 Cayman Islands Survey

Before 1924 the Cayman Islands (administratively managed by Jamaica) were not surveyed geologically. Between November and December 1922 an America oil prospector had visited those islands, and up to the time of Matley no report on that assessment was prepared. Matley was therefore deployed in 1923 to conduct a reconnaissance survey (Matley 1923c), and in January 1924 he was again posted there to conduct a more detailed survey for the Caymanian Government as approved by the Governor in Jamaica (Matley 1924c, d). In his preparation Matley found only two geographic studies for the islands dated 1912 (English 1912) and 1917 (Hydrick 1917). He spent 11 days surveying, and noted that the time spent was 'sufficient to make a fairly accurate geological map of the three islands as their geology is of a simple character'. Matley identified that the geological conditions combined with the small area of the islands restricted the development of good natural water supplies. He commented on road stone availability, phosphate deposits, and the potential for oil. He did not think that there was any petroleum potential based on what he had observed in Jamaica. However, he interpreted the islands as 'emerging peaks of a submarine ridge' possibly a dome structure which may form an oil trap. His geological theories were challenged when he discovered a pitch-like substance on the windward shore of Grand Cayman. He debated whether this was from an oil seep and so had it tested. It was found to be waste crude, probably dumped by a passing ship. Matley's report, *Reconnaissance Geological Survey of the Cayman Islands* was completed on April 30, 1924, on his journey back to England (Matley 1924d). It was printed in a supplement to the *Jamaica Gazette* (Matley 1924e). In addition to his official reports, Matley developed his findings from the Cayman surveys into numerous papers that were published in various research publications (Matley 1924f, 1925, 1926a, b). His research represented a boom in geological knowledge for the Cayman Islands as previously there was no comprehensive assessment of these territories.

3.1.5 Termination

The financial allocation to support the Geological Survey was significantly reduced to less than one third of that approved when the programme started. Matley was also to be terminated (Stockley 1925b). He proceeded on leave and departed Jamaica to Avonmouth on April 21, 1924, on the *S. S. Bayano*. He took the Jamaican rocks and fossil specimens and arranged to have others shipped to him for study and comparison with those in the British and other museums, and to get specialist advice for those he could not identify. Immediately before, Matley had just completed the initial stages of zoning the White Limestone, with assistance from his American colleagues in identifying the fossils. Precise zoning was intended but this was never accomplished. Matley arranged with Stockley to collaborate on producing the remaining maps and reports. For the rest of his life, Matley would revisit Jamaica three times (Robinson 1996).

Stockley, in mid-April, was charged to complete all the outstanding work of the survey. He had also commenced microscopic work on the Jamaican rocks and minerals where some 120 rock sections were prepared for examination under the microscope. He noted that only five of these rocks were described by Hill. A fortnight after Stockley was placed in charge he was also given his three-month notice of termination. Although he had to use the time to complete the reports the Government also permitted him to complete archival work at the Institute of Jamaica that he had initiated in his personal time. Stockley recognized that changes at the Institute had resulted in the geological collection of the 1859 Survey by Sawkins and his geologists were on the verge of becoming a 'heap of stones' having no labels or explanatory notes. He stated that when Sawkins departed from Jamaica the collection was reorganized and only few explanatory notecards were printed. He envisioned that the collection could become a standard of reference, and the centre of interest for West Indian geology attracting visitors with scientific curiosities and research students from across the world. At that time the famous large museums at South Kensington or that in Washington were the destinations of scholars wishing to compare specimens. Stockley however wanted to 'reverse the tide' (Stockley 1925b).

In his final report in January 1925 Stockley referred to the survey as the Government Geological Department, alluding to the formal recognition given to the activity (Stockley 1925a). Both Matley and Stockley had collaborated on the surveys, with Matley prematurely sent back to England. When the report was due Matley could not be reached as he had gone on a world tour with his wife (Robinson 1996). Stockley had to prepare it. The programme, as he reported, officially ended September 1924. The declining fiscal allotments, the ever-increasing demand for geological surveys by the authorities and the offensive comments from the impatient politicians regarding the progress of work by the survey seemingly frustrated Stockley. It led him to write

> No real Geological Survey of Jamaica has been contemplated for sixty years. In order that this may be done thoroughly it is essential for the geologist to live in the country under examination.

> This cannot be carried out efficiently by making fleeting traverses or hurriedly examining one locality after another. It can be carried out only by careful and prolonged investigation, by penetrating all valleys and gorges and by actually passing over the country on foot. (Stockley 1925a, p 12).

Further, he stated that

> The need [for,] and field of a geologist is not often known, but there are benefits derived from a proper geological survey which are generally disregarded by the commercially minded. (Stockley 1925a, p 12)

This snap geological survey in Jamaica between 1921 and 1924 was tantamount to a knee-jerk reaction of the Government to address diminishing water supplies. Though it lacked careful consideration it demonstrated the gap in understanding the intended desires of the administrators who commissioned the activity and the technical officers who implemented it. The geologists recognized that natural resource surveys, such as that for water, could not be carried out unless the basic geology was understood. The administrators, however, recognized that the need for water was immediate and did not have the convenience of time in which the geologists had hoped to operate. The geologists should have recognized that a geological survey organization is established to provide practical geoscience information to the nation as specified by its Government as national priority. National Geological Survey agencies are market-driven, and this survey in Jamaica was established to mitigate a crisis and not to support scientific curiosities. The administrators wanted immediate results and in this instance, it was not their intention to conduct a comprehensive survey. Stockley however acknowledged that Jamaica could have benefitted immensely if the survey programme was conducted systematically (Stockley 1925b). A complete survey he states

> ... broadens the outlook and lends colour to the immediate environment. (Stockley 1925a)

After working in Jamaica, Stockley was deployed in Africa where he got the opportunity to participate and lead in a systematic survey. Like Jamaica, the Zanzibar Protectorate was concerned about the water supply of the capital and sought to have a survey conducted. However, on the advice of E. J. Wayland, the Director of the Geological Survey of Uganda, a complete survey was to be undertaken, and this would include an investigation of the reputed gold occurrences at Pemba Island. Stockley was appointed as Assistant Geologist to the Government of Zanzibar and was based in Tanganyika Territory at the Geological Survey at Dodoma (FWHM 1929; Mineralogical Society 1929). Within his first year he had assembled a collection of fossils from which he was able to deduce the palaeontology of Zanzibar (Government of Zanzibar 1927). He thoroughly explored the various sources of water and identified that local rainfall was the ultimate source. The gold occurrence was insignificant, and he further showed that both Zanzibar and Pemba were poor in mineral resources. Despite his findings, his geological report (Stockley 1928) was considered valuable (FWHM 1929) and the Government of Zanzibar in November 1931 declared their pleasure by bestowing him a national award for his services (Anonymous 1931). Stockley went on to head the Geological Survey Department in Tanganyika. His tenure in the Caribbean coupled with his vast experience in other places in the world would serve him well. He visited Jamaica in 1949 to conduct hydrogeological surveys for the West Indies Sugar Company. Later, in March 1953 he took up appointment as Director of the Geological Survey in British Guiana. This job required him to travel for two months each year to work as consultant and advisor to the Geological Departments of other British West Indian islands where he provided oversight to their activities. When the modern Geological Survey Department in Jamaica was established in 1949, Stockley spent just over three weeks touring with the Director and the other geologists examining the mineral resources which included the new developments in bauxite, gypsum, and cement (Anonymous 1954).

3.2 Developments in the 1920s and 1930s

The short-lived Government-commissioned geological survey in Jamaica led by Charles Matley in the early 1920s represented a boom in geological knowledge for the island. This was coupled by significant research conducted by independent scholars. Of notable mention are the palaeontological investigations of Charles Taylor Trechmann, geologist from Durham University. He was a frequent visitor to Jamaica since 1909 taking special interest in the abundance of sedimentary rocks (Anonymous 1932a). In the early 1920s he produced many outstanding works such as those examining the relationship between the Cretaceous and Tertiary rocks (Trechmann 1922a), rudist bivalve fossils (Trechmann 1922b), the Yellow Limestone and its fossils (Trechmann 1923), the Richmond shale formation (Trechmann 1924a) and the Cretaceous limestone and its associated fossils (Trechmann 1924b). His fascination with fossils led him during his 1929 visit, to the Blue Mountains just short of the peak where he collected fossil shells. These he later donated to the British Museum after describing them in geological magazines (Trechmann 1929; Anonymous 1932a). On his February 1932 visit he examined the *Barrettia* fossil in rocks which he concluded as being the same age as English chalk. Trechmann had general interest in West Indian geology and on this visit, he had planned to examine the geology of St. Kitts, Barbados, and Trinidad among other islands which he considered as being geologically unique. After leaving Jamaica in March, he was on his way to Trinidad before sailing back to Southampton (Anonymous 1932b). Trechmann travelled to places such as South Africa, Central Africa and New Zealand, and he compared his Jamaican fossils with his collection from other locations (Anonymous 1932a). His publications have remained timeless references for students and researchers.

It was still recognized however, that Jamaica was not the subject of a complete and systematic geological survey. Eminent Cambridge University geologist, Cowper Reed, during a vacation trip to the island in February 1934 stated that Jamaica was far behind other countries with regards to up-to-date geological survey and he recommended a 'minute [or detailed] geological survey' of the whole island (Anonymous 1934a). He had hoped to conduct some geological reconnaissance during his vacation, but this was hampered by inclement weather. Reed reminded that most of the recent geology papers for Jamaica were published by Americans. He explained this noting that

> Jamaica geology gives the key to the whole structure of the Caribbean, and it wants working out minutely with modern knowledge and there is plenty to be done. (Anonymous 1934a)

Four years after Reed's recommendations for geological survey, Matley revisited Jamaica. He had visited in late 1926 where he stayed for six months working on his map of Kingston. On this visit in late 1938 he wanted to finalize his work in Kingston with the intention to present it to the Geological Society of London the next year. He also wanted to complete his survey of the geology of parts of Clarendon. Matley, 72 years old, also came to showcase and seek assistance to publish a giant manual on Jamaican geology accompanied by a geological map that he had authored (Anonymous 1938). In soliciting financial assistance from the Government, he professed that his book has information on the mineral and water resources. He stated that

> I have spent my life doing research work. I do not want all this heavy work to be wasted. (Anonymous 1938)

He further revealed a huge, red-covered volume identifying it as a complete geological survey of Haiti, compiled during the American administration of that island, slyly noting that

> I'm sure Jamaica is not behind Haiti – and you don't want it!. (Anonymous 1938)

On June 26, 1940, *The geology of the Kingston District, Jamaica* was read before the Geological Society of London (Matley 1940).

The Government, in 1946, published a smaller four-page booklet on the *Outline of the geology of the Kingston district of Jamaica* (Matley 1946) but printing of the accompanying one inch to the mile-coloured map was delayed because of World War II (Fig. 3.2) (Matley 1947). The geology map was constructed on the 'Pomeroy' topographic map, but the scale was reduced. The map represented 340 square miles spanning as far as Bog Walk in the west to Halberstadt in the east. Matley (assisted by Stockley) is the first geologist to conduct detailed mapping of a section of Jamaica using large-scale base maps (Robinson 1996).

The Government had sponsored Matley to visit Jamaica in late 1943, but on his way, he became sick and was hospitalized in Virginia. Upon recovery he came to Jamaica in February 1944 and stayed for four months. During this time, he finalized the contributions from local experts for his monograph on Kingston. Matley intended to return to Jamaica in September 1945 to complete fieldwork on his proposed giant volume on Jamaican geology. Unfortunately, he had a stroke that, along with other health issues, led to him being incapacitated. On July 8, 1947, Matley passed away at his home at Leamington, England, at the age of 81 years (Garwood 1948; Robinson 1996). His monograph titled *Geology and Physiography of the Kingston District, Jamaica*, still unfinished, was edited by his friend Frank Raw and finally published in 1951 along with the reduced scale geology map (Matley et al. 1951). Matley's maps, field notes, and photographs were donated to the Institute of Jamaica which further passed them on to the Geological Survey which was established in Jamaica in 1949 (Robinson 1996). His proposed manual on the geology of Jamaica was never published.

Matley exemplified a passion for research and the publication of scientific writings. He became a distinguished member of the Geological Society of London having been awarded the Murchison Medal along with a gift of ten guineas in 1928 for his work in the examination of stratigraphy of various parts of the British Empire (Gregory 1929). This honour, it seems motivated him in his career, and led him to cultivate a similar attitude of curiosity in young secondary school students whom he encouraged to explore natural history through the annual Matley Natural History Essay Competition which he founded. This was administered through the Institute of Jamaica (Anonymous 1943).

3.3 Mining Possibilities

While mineral exploration in Jamaica may have been minimal during the 1920s and 1930s, there was still some work done. In Castleton and Montpelier districts on the border of St. Mary and St. Andrew parishes D. C. Hylton was actively exploring copper–gold prospects (Siriunas 1993). Between 1931 and 1932 his party extracted and shipped a small load of ore for testing, but this proved uneconomical in processing. Even later in 1945, the old Clarendon copper–gold mines developed by the Spaniards were re-evaluated. The old shafts were discovered but attempts to restart mining failed (Hose 1950). During his visit in February 1934, Reed had also commented that 'gold find is possible in this Island' (Anonymous 1934a). Later in August it was announced that an American mining engineer and geologist Donald G. Miller was in the island to investigate 'gold mining possibilities' (Anonymous 1934b). He came for one and a half weeks to conduct 'special surveys for private parties' and investigate the potential for mining gold. He was careful to withhold the names of the parties to whom he was advising and the properties that he was interested in, declaring them a closely guarded secret. He however revealed that he had visited four areas with 'good' possibilities where at the surface he saw very good indications of minerals. Though Miller endorsed that the possibilities for finding gold in Jamaica was good, he commented further on the discovery success of the Spaniards. He emphasized that many mines in the Caribbean and Central and South America were discovered by the Spaniards. He noted

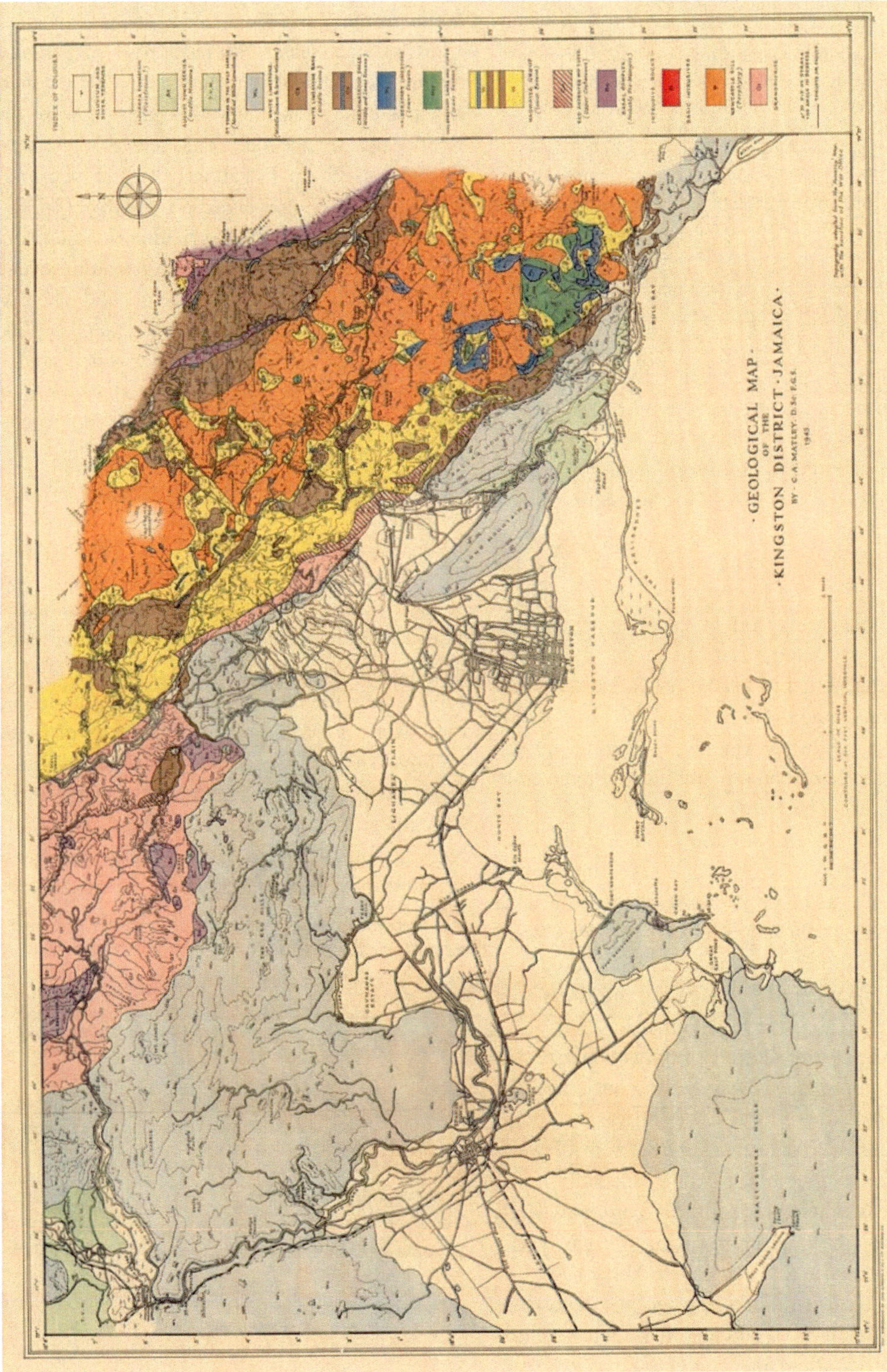

Fig. 3.2 Geological map dated 1945 prepared by Charles Alfred Matley (with Gordon Murray Stockley) that accompanied his *Geology and Physiography of the Kingston District* (Matley et al. 1951) This mapping effort is the first of its kind in Jamaica where a detailed geological map was prepared using large-scale base maps (Courtesy of the Mines and Geology Division, Jamaica)

> The Spaniards have been premier geologists in not only Jamaica but in other parts of the Caribbean and South and Central America. There are hardly any mines on Central and South America and the Caribbean that were not originally discovered by the Spaniards. That of course does not mean that these mines were successfully operated by them. Maybe that is why they allowed the British people to take Jamaica, and I can go further and say if the Spaniards had found gold, they would not have given Jamaica to England so easily. (Anonymous 1934a)

3.4 Red Infertile Soil

The abundant red soil draping the white limestone, especially in the central and some Western parts of the Jamaica was recognized by the Spaniards as a useful mineral component in building construction. The Spaniards mixed this terra rosa with slaked lime to make a reddish-brown mortar (Zans 1952). Henry Thomas de la Beche during his visits in the 1820s identified a 'red marly soil' with an upper part of 'rather compact red sandstone' occurring on the White Limestone in Manchester parish and other places in Jamaica (De la Beche 1827). Charles B. Brown while working with Sawkins in the mid-1800s, described this 'red ferruginous earth' found in Manchester as a soil of 'principally a mixture of iron and alumina' with 'minute concretionary pellets of iron and sometimes manganese' (Sawkins 1869). Brown noted the wide distribution of the deposits and that they were restricted to the limestone areas. He recognized too that the deposits varied in thickness and reached up to 10 m at Mile Gully.

Over the years the red soil was analysed for various purposes but was never considered for its metallic content. The material was still commonly used to make mortar in 'Spanish' walls or wattle and daub construction especially in the rural areas where the soil was common (Zans 1952). For many years the Kingston gas works mixed the terra rosa from Red Hills with saw dust and used it for filtering the gas, a mixture more efficient in removing hydrogen sulphide from the gas than the bog ore imported from Holland (Zans 1952). However, it was not until the late 1930s before the economic importance of this terra rosa as bauxite, the ore of aluminium, was recognized.

Most occurrences of terra rosa in Jamaica drapes the White Limestone and it was generally postulated that the soil originated from the weathering of this rock. This red soil being the largest soil group in Jamaica was of considerable importance and became the subject of scientific studies by the Agricultural Chemistry Division of the Department of Agriculture which sought to further understand the soil chemistry. It was long known that fertilizers applied to these soils was unusually short-lived, and it was suspected that this was linked to the clay content. Senior Agricultural Chemist, Robert Francis Innes directed the research along with H. F. Birch, Assistant Agricultural Chemist (Jones 1952). Soil investigations started in 1938 and focused on Grove Place, Manchester, west of Kendal on the road to Mile Gully, and at Bull Savannah, St. Elizabeth, and to lesser extent in other places. The studies highlighted the aluminium-rich minerals (Jones 1952). The economic potential however was not highlighted until April 7, 1942, when Innes presented the findings from tests conducted on soil samples from Lydford in St. Ann parish (Innes 1951). The material was then deemed as potentially bauxitic. Innes and his fellow researchers of the Agricultural Chemistry Laboratory are credited for designating some of the Jamaican terra rosa as a type of bauxite.

References

Anonymous (1921a) Colonial Secretary submitted annual estimates to council yesterday. Daily Gleaner, 4 Mar 1921, p 6

Anonymous (1921b) Letter from Dr. J. W. Evans. Daily Gleaner, 18 Mar 1921, p 1

Anonymous (1921c) Business transacted in the Island's Legislature yesterday. Daily Gleaner, 13 May 1921, p 3

Anonymous (1922a) Legislators continue the consideration of the estimates. Daily Gleaner, 17 May 1922, p 6

Anonymous (1922b) Work of the Island's Legislature yesterday. Daily Gleaner, 20 Jul 1922, p 5

Anonymous (1923a) The report of Dr Matley on his work on the progress of geological surveys in

Jamaica for economic purposes. Daily Gleaner, 5 Jul 1923, p 13
Anonymous (1923b) Hermitage dam discussed. Daily Gleaner, 5 Feb 1923, p 11
Anonymous (1923c) Honourable legislative council deals with estimate. Daily Gleaner, 27 Apr 1923, p 6
Anonymous (1923d) The autumn session of the hon. Legislative Council. Daily Gleaner, 21 Nov 1923, p 3
Anonymous (1924a) Legislators continue to deal with the estimates. Daily Gleaner, 30 Apr 1924, p 6
Anonymous (1924b) The Legislative Council. Daily Gleaner, 8 May 1924, p 14
Anonymous (1924c) Business brought before the Legislative Council yesterday. Daily Gleaner, 9 May 1924, p 6
Anonymous (1924d) Dr Matley's report on our water resources. Daily Gleaner, 30 May 1924, p 6
Anonymous (1931) Notices. The London Gazette, 24 Nov 1931, Issue 33774, p 7578
Anonymous (1932a) BR geologist here to study formations. Daily Gleaner, 23 Feb 1932, p 21
Anonymous (1932b) Departure record. Daily Gleaner, 31 Mar 1932, p 18
Anonymous (1934a) Jamaica geology gives key to whole Caribbean structure says English geologist. Daily Gleaner, 6 Feb 1934, p 6
Anonymous (1934b) American engineer of mines comes to investigate Gold Mining Possibilities here. Daily Gleaner, 20 Aug 1934, p 3
Anonymous (1938) Geological survey of Jamaica to be published if Government Help. Daily Gleaner, 13 Dec 1938, p 7
Anonymous (1943) The Matley natural history competition. Nat Hist Notes 1(10):6
Anonymous (1954) Considerable achievement seen in water supplies here. Daily Gleaner, 2 Feb 1954, p 5
De la Beche HT (1827) Remarks on the geology of Jamaica. Trans Geol Soc Lond series 2 2(2):171–173
English TMS (1912) Some notes on the natural history of Grand Cayman. In: Handbook of Jamaica
FWHM (1929) Reviewed work: "Report on the geology of the Zanzibar Protectorate by G. M. Stockley." J Royal African Society, 28(111)(April, 1929):320–321
Garwood EJ (1948) Obituary notices—Charles Alfred Matley. Q J Geol Soc 104:lviii–lx
Gregory JW (1929) Report of the Council for 1928. Q J Geol Soc 85:xi–xiii
Government of Zanzibar (1927) Report on the Palaeontology of the Zanzibar Protectorate, based mainly on the collection made by G. M. Stockley, government geologist, 1925–26. Government of Zanzibar, London
Hill RT (1899) The geology and physical geography of Jamaica: study of a type of Antillean development. Based upon surveys made for Alexander Agassiz. With an appendix on some Cretaceous and Eocene corals from Jamaica by T. Wayland Vaughan. Bull Mus Comp Zoo Harvard College 34
Hose HR (1950) The geology and mineral resources of Jamaica. Col Geol Min Res 1(1):11–36
Hydrick JL (1917) Report on hookworm survey of the Cayman Islands from April 18, 1917 to June 20, 1917 (The Rockefellar Foundation)
Innes RF (1951) Discovery of bauxite in Jamaica. Col Geol Min Res 2(3):270
Jones GA (1952) The truth about bauxite discovery and development in Jamaica. Daily Gleaner 20:8
Matley CA (1923a) Report on the work of the Government Geologist and the progress of the geological surveys in Jamaica for economic purposes. Supp Jam Gaz 46(9):225–231
Matley CA (1923b) Report on the hydrology of the Liguanea plain and on the possibilities for irrigating it. Jam Gaz (Extraordinary) 45:(10 August 1923)
Matley CA (1923c) Report of a Reconnaissance Geological Survey of the Cayman Islands. Geol Survey Cayman Islands, pp 41–45
Matley CA (1924a) The geological surveys here. Daily Gleaner, 25 Jul 1924, p 14
Matley CA (1924b) Report by the Government geologist on the progress of the geological surveys in Jamaica in connection with its water resources for the period June, 1923, to April, 1924. Supp Jam Gaz 47(7):45–48
Matley CA (1924c) Geological survey of the Cayman Islands. Annual general report, Jamaica, pp 41–45
Matley CA (1924d) Geological survey of the Cayman Islands. Daily Gleaner, 20 Jun 1924, p 13
Matley CA (1924e) Report of a Reconnaissance Geological Survey of the Cayman Islands. Supp Jam Gaz 13 Jun 47:69–73
Matley CA (1924f) Reconnaissance Geological Survey of Cayman Islands, B.W.I. The Pan-American Geologist XLII:313–315
Matley CA (1925) A Reconnaissance Geological Survey of the Cayman Islands, British West Indies. In: British Association for the Advancement of Science, Report of the 92nd Meeting. Toronto, pp 392–393
Matley CA (1926a) The geology of the Cayman Islands (British West Indies) and their relation to the Bartlett Trough. Q J Geol Soc Lond 82:352–387
Matley CA (1926b) Phosphate in the Cayman Islands. In: 14th International Geological Congress, Madrid, pp 777–779
Matley CA (1940) The Geology of the Kingston district, Jamaica (with discussions by C. T. Trechmann, and K. W. Earle). Proc Geol Soc Lond 1371:99–106
Matley CA (1946) Outline of the geology of the Kingston district of Jamaica (with one folded coloured map on the scale of one inch to one mile). Crown Agents of the Colonies for the Institute of Jamaica, London
Matley CA (1947) Geology of Kingston in booklet form. Daily Gleaner, 7 May 1947, p 16
Matley CA et al (1951) Geology and physiography of the Kingston district, Jamaica. Crown Agents of the Colonies for the Institute of Jamaica, London (edited by F Raw)
Mineralogical Society (1929) "Bye-Laws," [Mineralogical Society of Great Britain and Ireland]. Mineralogical Mag and J Mineralogical Soc 22(124, March 1929):viii–xxvi

Phillippo JC (1881) The Mineral Springs of Jamaica (Lecture VI delivered at the Town Hall, Kingston on the 3rd May, 1881). Henderson & Co., Kingston

Robinson E (1996) Charles Alfred Matley: his links with mid-twentieth century geology in Jamaica. Contributions to Geology, UWI, Mona 2:20–26

Sawkins JG (1869) Reports on the Geology of Jamaica, or Part II of the West Indian Survey, with contributions from G. P. Wall, Lucas Barrett, Arthur Lennox and C. B. Brown, and an Appendix by R. Etheridge. (Memoirs of the Geological Survey), Longmans, Green, and Co, London

Sawkins JG, Brown CB (1865) Geological map of Jamaica—scale: one mile to ¼ of an inch. Vincent Brook, Day & Son Lith., London

Siriunas JM (1993) Draft report no. 27, Inventory of Metallic Mineral Occurrences in Jamaica. Geological Survey Division, Kingston

Stockley GM (1925a) Final report of the Government Geological Department here. Daily Gleaner, 3 Apr 1925, p 12

Stockley GM (1925b) Final report of the Government Geological Department here. Daily Gleaner, 1 Apr 1925, np

Stockley GM (1928) Report on the Geology of the Zanzibar Protectorate, with a preface by E. J. Wayland, Director of the Geological Survey of Uganda. Government of Zanzibar, London

Trechmann CT (1922a) The Cretaceous and Tertiary question in Jamaica. Geol Mag 59:422–431

Trechmann CT (1922b) The Barrettia beds of Jamaica. Geol Mag 59:501–514

Trechmann CT (1923) The Yellow Limestone of Jamaica and its Mollusca. Geol Mag 60:337–367

Trechmann CT (1924a) The Carbonaceous Shale or Richmond Formation of Jamaica. Geol Mag 61:2–19

Trechmann CT (1924b) The Cretaceous limestones of Jamaica and their mollusca. Geol Mag 61:385–410

Trechmann CT (1929) Fossils from the Blue Mountains of Jamaica. Geol Mag 66:481–491

Wilson HE (1985) Down to Earth: one hundred and fifty years of British Geological Survey. Scottish Academic Press Ltd., Edinburgh

Woodring WP (1925) Miocene mollusks from Bowden, Jamaica; pelecypods and scaphopods. Carnegie Institute of Washington, Publication 366

Woodring WP (1928) Miocene mollusks from Bowden, Jamaica. Part II, gastropods and discussion of results. Carnegie Institute of Washington, Publication 385

Zans VA (1952) Bauxite resources of Jamaica and their development. Col Geol Min Res 3(4):307–333

4 The Bauxite Industry and Its Development

Abstract

The evolution of Jamaica's bauxite industry is explored. Jamaican bauxite was seemingly a subeconomic resource until innovative research upgraded it to a valuable commodity which provided great financial benefit to the country. The aluminium industry is global, highly political, and complex, and Jamaica's journey of 70 years of commercial bauxite mining is examined for the experience in navigating the changes. The successes and failures are described, and the past and present indeed provide insight into the future. The key lesson is that Government should create the enabling environment where the benefits from mining can be optimized and passed to their people to make them prosperous.

4.1 Bauxite and Its Importance

The metal aluminium was first discovered and named by British scientist, Sir Humphrey Davy, in 1808. French chemist, Pierre Berthier while working near Les Baux in southern France in 1821, discovered soil with high quantities of aluminium oxide (or alumina) and named it bauxite after the village (Davis 1989; Gow and Lozej 1993). The material was deemed a potential source of aluminium; however, a very high electric current was needed to isolate the metal. However, Friedrich Wohler of Germany, also in the 1820s, had managed to separate the metal from the oxide using a technique that would be extremely costly if scaled up. The exorbitant cost of the extractive metallurgy resulted in aluminium being a scarce and highly prized metal having value surpassing gold. However, Henri Sainte-Claire Deville of France refined Wohler's process in 1854 to develop an economical commercial-scale technique which led to aluminium prices declining over the next decade. This decline, however, had little impact and widespread use of the metal in industry remained limited (Davis 1989).

In 1873, the dynamo capable of producing hundreds of amperes of current was invented. This propelled the development, in 1887, of the Hall-Héroult process of electrolytic reduction of alumina. The process was developed at the same time but independently by American, Charles Martin Hall and Paul Louis Toussaint Héroult of France (Gow and Lozej 1993). The development of Hall's technique was funded by Captain Alfred E. Hunt. Therefore, the first application on a commercial scale occurred in 1888 at the Pittsburgh Reduction Company which was owned by Hunt. This company later renamed to Aluminum Company of America, or Alcoa. Alcoa in Scotland adopted the process in 1896. Héroult's technique was deployed in

S. Bhalai, *The History of Mining and Geological Surveys in Jamaica*, Historical Geography and Geosciences,
https://doi.org/10.1007/978-3-031-42604-9_4

Switzerland in 1889 by Aluminium Industrié, which became a part of a company called Alcan, and at British Aluminium, which was also connected to Alcan (Davis 1989).

Commercial exploitation of bauxite was revolutionized in 1889 when Austrian, Karl Josef Bayer, son of the founder of the Bayer Chemical Company, invented the Bayer process for the large-scale extraction of alumina from bauxite. Hot caustic soda was used to leach the bauxite to remove the alumina content. France was the leading producer of bauxite followed by the USA. Both countries supplied the world until World War I. At the beginning of the twentieth century world bauxite production was less than 500,000 tons per year but by 1914 when the war began it nearly doubled (St. James Press 2000). Pittsburgh Reduction in 1899 had created a subsidiary in Canada having of a power plant and reduction facility at Shawinigan Falls, Quebec. Aluminium production started in 1901 and by 1902 the subsidiary became known as Northern Aluminum Company Ltd (St. James Press 2000). The main usage for the metal in peace time was in cooking utensils. Aircraft technology was still undeveloped, but the demand for aluminium was increasing.

During the First World War, high-grade bauxite deposits were identified in Guyana and Suriname (Tretzel 1971). The Guyanese deposits were immediately developed in 1917; however, extraction of the Surinamese deposits started later in 1922. Production in Italy, Hungary, Yugoslavia, and Russia also came online, but France remained the main supplier in Europe. In Asia, deposits were found in Indonesia and Malaysia. New applications for aluminium were being developed including extrusions and foil; applications that formerly utilized tin and lead. The United Foil Company of Louisville, Kentucky, founded in December 1919 by Richard Reynolds was primarily engaged in producing lead and tin foil for packaging tobacco. Richard's uncle, R. T. Reynolds, had created this niche industry. However, United Foil started producing aluminium foil in 1926, favouring it's more pleasant and shiny appearance to the other metals. The company changed its name to Reynolds Metals Company in 1928 (Davis 2000).

Before the Second World War the annual demand for bauxite was less than 5,000,000 tons, with France the leading world producer, followed by the USA, Guyana and Suriname. Alcoa monopolized aluminium production and marketing. Its subsidiary Northern Aluminum, in 1925, changed its name to Alcan, the shortened form of Aluminum Company of Canada, Ltd. By June 1928 Alcoa, the world's leading aluminium producer divested its foreign operations, forming a Canadian holding company called Aluminium Ltd. Shareholders received one stock of the new company for three shares in Alcoa. Aluminium Ltd was strategically set up to penetrate foreign markets and participate in international cartels, away from the scrutiny of the US Justice Department. This latter purpose was illegal for US companies (Davis 1989). Alcoa would continue to dominate the USA market. During the Second World War bauxite demand increased sharply and annual production was approximately 14,000,000 tons. The Germans actively monitored the shipping lanes using submarines to attack the ore ships. Shipments from Guyana and Suriname to the USA and Canada were restricted. This forced an increase and peaking in production of the US deposits which eventually led to early depletion of their high-grade ore (Davis 1989).

After the war the technological applications of aluminium increased which led to a corresponding increase in world demand and bauxite production. New mines were opened in Haiti, Ghana, Sierra Leone, Brazil, Dominican Republic, Jamaica, Australia, and Guinea (Tretzel 1971). The US Government, to break the monopoly of Alcoa, strategically sold their American aluminium facilities to Kaiser and Reynolds. The Government further provided loans on favourable terms that enabled these companies to maintain their market share and remain profitable while competing against the giant Alcoa (Davis 1989). In 1950 the US Government forced Alcoa to segregate itself from the foreign giant Aluminium Ltd on the basis that Alcoa and its allies had colluded to restrain imports and preserve its US monopoly. This case started in 1937 before the war, and it represented a landmark ruling where for the first time in history US investors were ordered by their Government to give up control in a foreign company. The case remained open for court supervision until 1957 when it was finally closed after a request by the Justice Department for extended monitoring was denied. Nine stockholders controlled nearly 45% of Aluminium Ltd stock and 46% of stock in Alcoa. All except Edward K. Davis, head of Alcoa, sold their stock in Alcan. Davis was the brother of Arthur Vining Davis, former head of Alcoa for whom, the Alcan town Arvida in Quebec was named (St. James Press 2000).

4.2 Bauxite in Jamaica

In 1942, four years after the discovery of bauxite in Jamaica, 69-year-old farmer, distinguished business leader and Member of the Privy Council, Sir Alfred Horace D'Costa observed that his plants were not thriving in the red soil on his property at Lydford, St. Ann. D'Costa experienced difficulty in establishing Wynne Grass (Molasses Grass) and growing corn. The Agriculture Department had earlier recognized that the infertility was caused by the low levels of phosphates and potash available to plants, and this was the typical advice provided to the landowner. D'Costa, however, requested that Government soil scientist Robert Innes test the soils from his Lydford, Belmont, and Crescent Park properties. The analyses showed the soils to be laden with alumina and iron but low in silica which Innes advised could be considered as low-grade bauxite. He further guided D'Costa that the findings would be of interest to the Imperial Institute of the Colonial Office in London which would assess the economic potential of the material (Jones 1952). The Imperial Institute, founded in 1883, was mandated to assess Commonwealth mineral resources and promoted the development and industrial use of raw materials in the British Empire during the first half of the 1900s. This specialized function resulted in the Geological Survey in Britain making limited contributions to economic geology at that time (Wilson 1985). The Director of Agriculture in Jamaica, G. A. Jones, contacted the Institute which requested economic information on the deposits and their respective tonnages. Innes prepared the report which the Institute in turn relayed to the Ministry of Aircraft Production in Britain. This Ministry was responsible for all shipment of such materials into the UK (Jones 1952).

D'Costa simultaneously submitted the findings to aluminium producers in New York and London, and the Canadian Government at Ottawa to stimulate their interests. The USA and British were slow to respond to D'Costa; however, the Minister of Mines in Ottawa referred the analyses and descriptions to the Department of Munitions and Supply and then to Aluminium Ltd at Montreal (which later renamed to Alcan). Aluminium Ltd requested detailed sampling of the soils along with other information that could be provided. D'Costa arranged with Innes, and the samples were dispatched to Ottawa at the end of May 1942. By July, analyses at Aluminium Ltd's Arvida Research Laboratories confirmed that the specimens were indeed a kind of low-grade bauxite. However, the quality was not immediately suitable for the North American alumina plants. Despite this, Aluminium Ltd recognized a potential commercial opportunity and immediately deployed Horace R. Hose of the subsidiary Aluminium Laboratories Ltd, Montreal, to visit Jamaica to examine D'Costa's properties and conduct other reconnaissance prospecting. At the time there was no Geological Survey agency in Jamaica to consult (Zans 1952).

Hose arrived in August 1942 and conferred with Innes towards surveying D'Costa's lands and other similar properties in St. Ann. His reconnaissance extended into other parishes where he concluded that Manchester, St. Elizabeth, and St. Ann hosted the more important bauxite deposits. Hose returned to Jamaica in November 1942 to advise His Excellency, the Governor of Jamaica, Sir Arthur

Richards of his positive findings and to arrange for Government authorization to conduct detailed exploration (Jones 1952). After meeting with Hose the Governor declared on November 27, 1942 under the Defence (Amendment No. 21), Emergency (Defence) Acts 1939–1940, Regulation 1942 that

> . . . any bauxite or minerals worked under this regulation shall on severance become the property of His Majesty and may be used or dealt with in such way as the Governor may think fit. (Zans 1952)

Aluminium Laboratories Ltd was appointed as the agent of the Government to conduct further investigation and to develop and mine the ore should it prove advisable to supplement production of bauxite from Guyana and elsewhere as a war emergency. The Government further allowed that company to use the Agricultural Department Experimental Station at Grove Place for their field and office base and placed the Agricultural Chemistry Laboratory of the Department of Agriculture at Hope, in Kingston, at the disposal of the company. Extensive exploratory work on the deposits was conducted, and thousands of samples were tested. This detailed survey proved the existence in several places, of bauxite having commercial possibilities. Aluminium Laboratories later established its own testing facility at Mandeville, and by February 1943 the Government gave the company permission to mine and ship 2,500 tons of ore from the Grove Place station property. This was sent to Alcoa's alumina plant at East St. Louis, Illinois, for experimental plant-scale investigations (Davis 1989). This represented the first shipment of Jamaica bauxite.

The experiment showed that Jamaican ore comprised a complex mix of minerals and could not be treated economically in the existing technology at the North American alumina plants which employed the conventional Bayer process. Treatment of Jamaican ore would require modification of equipment and the extractive technique or would result in considerable sacrifice of production capacity (Davis 1989). This necessitated further pilot studies.

During this time, shipments of bauxite from Guyana were still under threat by the Germans who had already sunk several South American cargoes. However, by mid-1943 the attacks subsided and so did the necessity to develop Jamaican bauxite as a strategic resource. The emergency Defence Regulation governing the mining of bauxite was gradually withdrawn in 1944 and was not replaced with any permanent law. In October 1943 Aluminium Laboratories Ltd established Jamaica Bauxites Ltd as a fully owned subsidiary to manage their exploration activity in the Island. The outfit was based in Mandeville and began acquiring bauxite-bearing properties early in 1944, which by 1946 the land prices had risen by two to three times their former value (Zans 1952).

The shift of Aluminium Laboratories Ltd into Jamaica led to other companies also seeking commercial opportunities for bauxite mining. The Dutch, N. V. Billiton Maatschappij had also obtained options on properties in St. Ann in 1943. This company was involved in tin mining in Malaysia and bauxite mining in South America. After their assessment of Jamaica, the company withdrew in the same year (Zans 1952). Reynolds Metal Company also expressed interest in Jamaican bauxite in that same year as part of the US Government's strategy to break the dominance of Alcoa. Earlier in August 1940, US President F. D. Roosevelt had called for a supply of 50,000 military aircrafts per year. He soon created the National Defence Advisory Commission (NDAC) to direct production. Investments of US $739 million were made to build two alumina plants that would be supplied with Jamaican bauxite, nine smelters, and several fabrication plants. Alcoa, formerly Pittsburgh Reduction, however, still had the patent for reducing alumina and also dominated aluminium production in the USA. Therefore, nearly all the plants installed by the US Government were operated by Alcoa. Reynolds attempted to build a smelter by accessing funds from the NDAC but was unsuccessful. With the help of Senator Lister Hill of Alabama, Reynolds finally got a US $20 million loan in

August 1940 to construct smelters at Listerhill, Alabama (in appreciation to the senator), and Longview, Washington (Davis 1989).

After the war the US Government also disposed of its surplus assets and prospective competitors of Alcoa were given preference of plant and equipment. Alcoa was given opportunities to certain facilities but it was subject to the approval of the Attorney General, and the acquisitions should not give Alcoa any competitive advantage. Both Reynolds and Kaiser Aluminum and Chemical Corporation got favourable lease agreements and purchase options. Reynolds acquired the Hurricane Creek alumina plant in Arkansas, the Listerhill sheet mill, and a smelter at Troutdale, Oregon. It also obtained the rights to purchase bauxite from the Government's strategic stockpile for the Hurricane Creek plant and the right to use Alcoa's patent royalty-free at that plant (Davis 2000).

Reynolds, however, still lacked the steady bauxite supply for its plant. Alcoa controlled deposits in Suriname, while Aluminium Ltd had preference in Guyana. Reynolds' vice president Walter Rice in November 1942 wrote the US Secretary of State, Cordell Hall, to obtain the State Department's support to obtain access to Jamaican ore. He noted that Jamaican bauxite would be of strategic interest to the USA, citing that the British Empire, with the large Jamaican deposits and high-capacity smelter at Saguenay River, Canada, would be able to dominate the international industry for 100 years and stifle the growth of the US aluminium industry. He stressed the importance of Jamaican ore to the 'industrial future' and 'national security' of the USA. Reynolds then, in January 1943, communicated with the US Bureau of Mines to obtain information on Jamaican bauxite. Rice obtained some notes on the location, size, and chemistry of some deposits. Communication was also made to John Huggins, who later became Governor in Jamaica. Huggins at that time was attached to the Anglo-American Commission in Washington. Reynolds requested the co-operation of the Jamaican Government in exploring, acquiring, and mining the ore (Davis 1989).

Both Rice and the chief geologist of Reynolds, Otto Carl Schmedeman, visited Jamaica in February 1943 to meet with Governor Richards. They sought similar terms like Alcan and Billiton, justifying their request based on meeting the quota of aluminium needed for the aeroplane production call made by President Roosevelt (Davis 1989). Governor Richards supported Reynolds and wrote on their behalf to the British Secretary of State for the colonies in March 1943. He cited that it would not be in Jamaica's best interest to grant Alcan the concession for the whole island. The US consular and other officials also showed their support. The Secretary of State for the colonies referred the matter to the Combined Raw Materials Board which did not approve. This board was set up by the British and American Governments to coordinate the procurement and allocation of raw materials following the December 1941 attack on Pearl Harbour by the Japanese. The board had an Aluminium Committee which awkwardly had members connected to Alcan. One such member belonged to British Aluminium, an ally of Alcan, and another was the Canadian Metals Controller (Davis 1989).

By May 1943 the US Government communicated to Reynolds that the earlier decision of the Aluminium Committee against the company exploring and mining bauxite in Jamaica was withdrawn, and they were free to advance their plans. The company also learnt that Alcan's pilot plant studies done in Illinois had failed as the existing plants were not outfitted to process Jamaican ore. The inability to process ore and the war shifting favourably to the allies reduced the demand for Jamaican bauxite and led to the emergency regulation being repealed (Davis 1989).

By late 1944 Reynolds started acquiring bauxite-bearing lands and began active exploration. Initially, companies were required to acquire ownership of the bauxite lands. By November, 15,000 acres were acquired in St. Ann, Manchester, and St. Elizabeth, and this soon increased to 60,000 acres. The company was also granted use of the Agricultural

Chemistry Laboratory. Company geologist and vice president of Reynolds, Schmedeman, after examining the bauxite deposits recommended the large-scale development of the Jamaican resources. The company also undertook pilot plant tests in the USA to develop a modified processing technique for the ore. Both Schmedeman of Reynolds and Hose from Alcan spearheaded the pioneering works on the geology of Jamaican bauxite. During this time, several small cargoes of ore (about two or three in 1948) were shipped for experimental purposes, but commercial mining was not yet established (Zans 1952). The small particle size, the mixture of aluminium-bearing minerals, and the high phosphorous content were the three main problems that affected the efficiency of extracting alumina (Chin 1971).

Reynolds, upon solving the processing challenges, moved operations to Jamaica to begin mining. This was further encouraged by increased usage of aluminium in industry, favourable taxation agreements between Jamaica and the USA, and the demands arising from the Korean War. The USA, ally to South Korea, needed to replenish its aluminium stockpile. This led to Reynolds constructing an alumina plant at Corpus Christie, Texas, to use Jamaican ore. The bauxite was won from mines at Lydford in St. Ann parish, transported 8 miles to the north, and exported from the Ocho Rios Port to Texas. On June 5, 1952, the first commercial shipment of ore was exported (Davis 1989). Schmedeman remained as vice president at Reynolds Jamaica Mines Ltd until his death in early December 1951 in Mexico City where he was deployed to negotiate for some fluorspar mines. He also held position as vice president for the Reynolds Mining Corporation and chief geologist for Reynolds Metals Company (Buchholz 1952).

The entry of Reynolds to Jamaica was followed by Permanente Metals Corporation, subsidiary of Kaiser Aluminum and Chemical Corporation, which started prospecting and land acquisitions in 1947. The company identified sufficient material in St. Elizabeth and Manchester to undertake operations similar to Reynolds where ore was mined and shipped to processing plants on the US Gulf Coast (Zans 1952).

As for Aluminium Ltd, on completion of prospecting work by subsidiary Jamaica Bauxites Ltd, Hose was replaced, and D. A. Bryn Davies took over in 1946 as Managing Director and geologist responsible for the development of the bauxite deposits. Hose departed in June for Canada before going to Malaysia to resume bauxite exploration work that was interrupted by the war (Anonymous 1946). Originally, Aluminium Ltd intended to mine and ship some of the Jamaican ore in crude form and process some locally to alumina, however, instead all the ore was processed locally.

4.3 Mining Laws and Policies

The seventeenth century Proclamations of Oliver Cromwell (Cromwell 1655), and that of Charles II (Charles II 1661) which granted land and mineral rights remained in force until the 1940s. In common law, the grant of land in fee simple carried the grant of all minerals except gold and silver unless otherwise reserved. There was no update to the rules governing mining in Jamaica (Hose 1950), and this gap may reflect the insignificant value of mining in the economy of the colony. It was thought that the Spanish had exhausted the more valuable minerals such as gold and copper, and the colony no longer had any resource of substantial value. Further, the Government may have avoided any real development of minerals, if there indeed was any, to avoid potential disasters similar to the economic failures of the mining ventures in the 1800s (Huggins 1965).

In the 1940s, the red soils distributed across the limestone districts of the island had become economically important. Foreseeing the potential for economic gains from the abundant bauxite resources, it necessitated that Jamaica documented in detail the terms and conditions for mining. Special attention was needed for the rights to the minerals, the length of the mining leases, and the income such as royalty, taxes, and customs duty to the island. In 1943 Captain Frederick Burnett, Commissioner of Lands explained to the Colonial Secretary the need to regulate mines as Aluminium Laboratories was acquiring bauxite lands cheaply. He also warned that landowners may hike the prices when they find out their land contains minerals. To prevent speculation, he recommended that the Government should pass a law to vest minerals in the Crown with provisions for owners to receive a compensation for the material extracted. By October that year the Secretary of State for the Colonies informed the Governor that orderly exploitation of minerals would require vesting the rights in the Crown. He stated, however, that the Legislature at the time was unlikely to pass a suitable law, and as such, that position should be placed on hold until the new Legislature was in place. At the time, Jamaica was in the process of achieving limited self-government and under the new Constitution, a House of Representatives was elected (Davis 1989).

By November 1943 the Attorney General, Henry Myers, summarized for the Colonial Secretary the legal position for minerals in Jamaica which addressed the common law grants of land and mineral rights and the separate treatment for petroleum. He recognized that the public was now aware of the bauxite land possibilities, and this may stimulate varying responses. He therefore recommended that the controls under the Defence Regulations of November 1942 be retained until a decision was reached. Further, he advised that both the public and the bauxite companies should be informed (or in fact, warned) that mineral rights and corresponding legislation was being considered and might be enacted retroactively. The Governor, however, communicated to the Secretary of State for the Colonies that vesting in the Crown could lead to landowners being entitled to enormous compensation, and it would be useful to be guided by the experience of other colonies. He later rescinded the vesting amendments of the Defence Regulations in December 1944.

On February 16, 1945, the Executive Council instructed that a bill should be drafted for the vesting of minerals in the Crown and compensation to land owners as a percentage of royalty should be included. Both Aluminium Laboratories and Reynolds were already advised that mining legislation was to be prepared and enactment may be retroactive. The Government obtained R. S. McKilligan, an Inspector of Mines from Trinidad, to provide technical guidance in the preparation of the law. The Jamaican Government would contribute to his salary for this special assignment (Government of Trinidad and Tobago 1949). By March 29, 1945, McKilligan reported to the Colonial Secretary specific considerations from his technical assessment of Jamaican bauxite deposits which he compared with the Guyanese deposits to determine which factors to address. He indicated that the 1930 Bauxite Mining Regulations of Guyana could guide the drafting of the Jamaican laws.

On July 17, 1945, the draft Vesting Bill was sent by the Governor Sir John Huggins, to the Secretary of State for the Colonies, Hon. G. H. Hall. Hall had just been installed following the British Election of July 5, 1945. In October 1945, he directed the Governor that mining legislation should be separated from the minerals vesting bill. He also stated too that a reply to Aluminium Laboratories' letter on minerals policy was forthcoming. On May 17, 1945, Aluminium Laboratories had written to the Secretary of State for the Colonies, Colonel Oliver Stanley, requesting consideration for an aluminium industry policy for the British Empire. The company had operations in Guyana with established terms and conditions. However, there was uncertainty of what was to be done in Jamaica. The company encouraged their idea citing the existence of their value chain and hence the channelling of company investments in the Commonwealth countries. They mentioned the abundance of good quality bauxite in the West Indies, West Africa, India, and Malaysia, hydroelectricity in Canada for smelting, and that the reliable source of skilled labour of the UK can be used for the fabrication of products to supply the domestic and foreign markets. The reply letter was dated November 1945.

On August 17, 1946, the new Secretary of State for the Colonies, Arthur Creech Jones (who replaced Hall) provided the Jamaican Governor with a Memorandum on Colonial Mining Policies. All Colonial Governments had received copies to guide the preparation of their own policy. This lengthy document was also published in London on December 11, 1946 (Anonymous 1946) and stated that

> In view of the importance on mining in the economic development of Colonial territories and the impact of mining operations on the social structure of the communities concerned, the Secretary of State considered it desirable that Colonial Governments should be furnished with guidance on the general principles which should be followed in framing mining policy. (Anonymous 1946)

The Memorandum identified that

> Minerals are important economic assets to a territory and being a gift of nature, their benefits should be shared by the community generally to which they belong, and not to be enjoyed merely by limited groups of private individuals who are often not members of the community concerned. (Anonymous 1946)

The memo recommended that mines should be efficient, consider the protection of other natural resources, and entail good standards of labour conditions and the proper provisions of welfare and social arrangements. It also noted that provisions should be made for indigenous populations to fit themselves for the highest technical and administrative posts, and importantly, an adequate share of the proceeds of mining should be retained in the Colony (Anonymous 1946).

In summary the purpose for a mining policy as recommended by the Memorandum should

> . . . ensure that the greatest benefits to be derived from future discoveries of mineral wealth shall be earmarked for the Colonial people in whose territory they happen to be found. (Anonymous 1946)

By 1947 the two key draft mining legislations were ready for enactment: the Minerals (Vesting) Act and the Mining Act. The former was endorsed by the Governor on August 26, 1947, for effect on August 30, 1947. The latter was assented on September 7, 1947, and proclaimed by him on October 4, for effect on October 13. The Governor also named Captain Arthur Frederick Thelwell, Commissioner of Lands as the Acting Commissioner of Mines on October 6, 1947.

The duties of administering the mining laws, however, were steadily increasing with more mineral exploration for bauxite, gypsum, guano, cement raw materials, iron ore, silica sand, and other minerals. This was noticeable between 1949 and 1952 as the mining portfolio required more commitment from its chief officer, the Commissioner. The traditional role of the Commissioner of Lands involved administering the over 262,000 acres of Crown Lands together with all Government-owned property in the island, the island-wide Land Settlement Programme which included settlement of ex-service men from both World Wars, the administration of the Twickenham Park Agricultural Centre where new crop trial experiments and pilot-scale crop production were being undertaken, among many other responsibilities (Chubb 1961). To lessen the workload, it was considered to delegate the mining matters to the newly established Geological Survey Department. This department was formed in late 1949 as a seven-year programme under the Lands Department. It would include a Mines Division and one of the Government Geologists would be recommended to serve as the Commissioner of Mines.

All applications for mineral exploration were initially submitted to the Geological Survey where the Director, Verners Zans, provided recommendations on whether permits were to be granted or denied. The volume of applications was increasing, probably triggered by the release of the first bulletin of that Department on the *Economic Geology and Mineral Resources of Jamaica* (Zans 1951). This publication provided a comprehensive summary of the various mineral deposits in Jamaica, a compendium only comparable to the work of the first Geological Survey (Sawkins 1869) eighty years before. To regulate mines and ensure compliance the responsible officer had to become familiar with the new mining laws. The Director soon became overwhelmed, and the duties were further delegated to another geologist, Howard Versey. Both geologists recognized that the new duties were time-consuming and conflicted with the geological mapping mandate of the Geological Survey. Further, the survey was scheduled to end by March 1956 and the assumption of the additional duties presented a challenge to timely completion of the mapping programme and the proper evaluation of the mineral resources. The geologists asserted the challenge they faced and that led to the arrangement being abandoned (Chubb 1961).

The Government in 1952 instead proceeded to establish a Mines Division at the Lands Department at 124 East Street, Kingston. The Commissioner of Lands, who at that time was Bahamian national Caesar Peter deFreitas, retained the role as Commissioner of Mines; however, there was now a deputy and two Inspectors of Mines (Cover 1953; Government of Jamaica 1953). This division later became an independent Department of Mines on January 15, 1957, with W. K. Brown appointed as Commissioner of Mines supported by seven other staff members. In that year the department moved to leased premises based at 4 Constant Spring Road (Government of Jamaica 1959). It administered all laws and regulations in respect of the control of mines, quarries, and oil exploration governed by the Mining Act, the Minerals (Vesting) Act, the Quarries Act, 1955 (and later the Quarries Regulations passed on July 22, 1958), and the Petroleum (Production) Act, 1940. By 1959 plans were being drawn for the construction of permanent offices at the Government property at Hope (Chubb 1961; Government Public Relations Office 1961; Jamaica Information Service 1965).

4.4 The Companies

The world bauxite industry was initially dominated by the transnational corporations. Between the early 1950s and 1970s Alcoa operated in 26 countries other than the USA. Kaiser, Reynolds, and Alcan were positioned in 12 to 15 countries. The American companies extended first to Guyana and Suriname then later to Jamaica and Guinea, then Australia. The European companies, the British and Swiss also went to Australia (Richardson 1975).

Jamaica's bauxite resources were recognized in 1939 and with focused development work in overcoming processing challenges, commercial bauxite mining started in Jamaica in June 1952. By 1958 Jamaica was the world's leading bauxite producer (Roberts 1971). Just three years before, bauxite was also discovered in Australia near Weipa on the Cape York Peninsula, and mine development and shipping of ore closely followed. Deposits were also found in Western Australia. Production and development increased, and by 1968 Australia was fourth place in global production with reserves estimated to run in billions of tons only exceeded by Jamaica, Suriname, and Russia. At this time the deposits in Guinea were quantified to be over a billion tons. World reserves were estimated at 5.8 billion tonnes with the potential for additional resources of 9.6 billion tons (Tretzel 1971).

The time for recovery from the Second World War along with technological problems in processing Jamaican bauxite in the conventional Bayer alumina plants resulted in the stalled development of the deposits. The material had chemical and physical properties which limited its treatment in the North American processing plants. The standard Bayer process could not be applied without some modifications. Zans (1952) described the Jamaican bauxite industry as an example of 'commercial development of relatively low-grade, iron-rich bauxite that posed problems of varying natures'. It shows how large quantities of low-grade ore can be rapidly developed under favourable conditions and how modern mining schemes can be established prosperously. A material of previously no real value became economically important in just a few years. Both Aluminium Laboratories Ltd of Canada and Reynolds Metal Company spearheaded the research to find the necessary modifications to the Bayer process. The acquisition of mining concessions by the subsidiaries of the companies had begun as early as 1943 with Aluminium Ltd, followed by Reynolds in 1944, and Kaiser Aluminum and Chemical Corporation in 1947 (Roberts 1971). The difficulties were overcome, and the first commercial shipment of bauxite was sent off in mid-1952 by mining subsidiary Reynolds Jamaica Mines Ltd. Kaiser Bauxite Company made their first shipment in February of the following year (Roberts 1971). It is estimated that from 1942 when it was recognized that substantial quantities of bauxite existed, up to 1952 when commercial mining started, J$20 million had been expended on exploration and pre-production works to operationalize the industry (not including the capital for the companies processing plants). In 1952, only 340,000 tons of ore was mined, but by 1953, the first full year of production was accomplished with 1,154,172 dried long tons mined and 1,054,978 tons exported. Of the bauxite processed locally 31,460 tons of alumina were produced, of which 28,731 tons were exported (Beard 1954; Hill 1971; Walker 1971).

Jamaican bauxite deposits varied in size from less than 50,000 tons found on an acre to over 3 million tons over an area greater than 76 acres. However, the average deposits ranged from 250,000 ton to 400,000 ton spread across 5–12 acres and reaching depths of 24 to 36 ft. Bauxite mining companies were required to obtain a mining lease to access the reserves, and they also needed to own the lands. The latter resulted in random land ownership across the island and the companies being the major landowners. However, an amendment to the mining laws in 1957 resulted in companies being issued with 25-year-long mining leases, and land ownership was not mandatory to obtain the lease (Roberts 1971).

Jamaica Bauxites Ltd had been renamed Alumina Jamaica Ltd since October 27, 1952, which later renamed Alcan Jamaica Ltd when the name Alcan was adopted by the parent company. Alumina Jamaica had originally planned to ship some bauxite and process some locally into alumina. The company instead processed all the ore locally at the new alumina pilot plant at their Kirkvine Works at Shooters Hill, Manchester. This was the only company locally processing bauxite to alumina, and this was the first alumina plant in the Caribbean. Construction started in November 1950. The company operated in Manchester, St. Ann, and St. Catherine parishes. It had staked a large claim on a 12-mile-long continuous deposit at Williamsfield to Mile Gully in Manchester. The Kirkvine Works alumina plant was situated at the end of this valley near Shooters Hill. Alumina Jamaica had set an industry milestone being the first company to build an alumina plant overseas. Their project investment totalled £13.5 million, which also included the agricultural schemes of the company (Zans 1952; Jamaica Information Service 1965).

The bauxite companies were all required to keep productive the lands they acquired until it was ready for mining. Alumina Jamaica acquired 30,000 acres, but only 12% contained commercial bauxite. The rest was for agriculture and afforestation. Alcan established an agricultural division to manage its lands which included research on how the land could be utilized before and after mining. The law required that rehabilitated lands to be restored with not less than six inches of topsoil, but the company scientists found that this was too thin for successful plant growth and therefore increased its limit to 12 inches. The company planted its rehabilitated lands with grass and undertook a very successful dairy operation (Zans 1952).

General contractors Sprostons (Jamaica) Ltd was contracted to build the Kirkvine plant which produced its first alumina by December 1952. It was designed to produce 100 tonnes of alumina per day but was soon expanded to 180 tonnes per day. By the end of 1953 it had expanded to 450 tonnes per day, and there were plans of reaching 600 tonnes per day. The plant was connected to the Jamaica Government Railway by a 1.5 km branch to the Williamsfield Station. The first shipment of alumina amounting to 2,300 tonnes left Jamaica early January 1953 to a smelter owned by Aardal og Sunndal Verk, Norway's largest aluminium producer at that time. In 1952 the company had mined 100,271 tonnes of bauxite, most of which was gibbsite or trihydrate ore having between 50% and 61% alumina. During that year the ore was held in

100,000-tonne emergency stockpiles. Ore was won from open pits and loaded by dragline excavators into wagons that transported it to the storage pit at the plant. The empty ore pits were used to store the red mud residue generated by the alumina plant (Zans 1952).

The calcined alumina produced at Shooters Hill was packaged in 100-pound bags, transported by the railway to the port at Kingston where it was shipped. This arrangement was temporary as a port was being constructed by Marples, Ridgeway and Partners Ltd at Old Harbour Bay, 56 km from Kirkvine. The work involved dredging a 2,134-m-long, 122-m-wide channel in the bay by removing 1.7 million cubic metres of sediments. This material was used in levelling the site and raising it well above sea level. It was completed in 1953 and named Port Esquivel (Zans 1952). Although Norway still received some alumina, by 1954 alumina was being shipped to the new company smelter at Kitimat, British Colombia. Kitimat represented the world's largest aluminium smelter and power development (hydroelectricity) at that time. This provided some encouragement to expand operation and process more ore locally. Therefore, by mid-1956 construction of a second alumina plant, Ewarton Works in St. Catherine began, and the first alumina was produced in October 1959. In this year Alumina Jamaica changed its name to Alcan, the shortened form of its parent Aluminum Company of Canada. With the construction of the Ewarton plant, Alcan would have invested £40 million in Jamaica. Since their construction up to 1968 both the Kirkvine and Ewarton alumina plants underwent a series of capacity upgrades which brought them each to 550,000 tonnes per annum. All the alumina produced at the plants were transported by the Jamaica Railway Corporation to Alcan's dock at Port Esquivel, located five kilometres west of Old Harbour Bay on the south coast of the island. The alumina was shipped to Canada, Scandinavia, and various other countries (Government Public Relations Office 1960; Jamaica Information Service 1965; Barker 1971).

Reynolds Jamaica Mines Ltd would invest US $17.5 million in their project. Construction of their facilities began May 1952 and mining started in June. The company had facilities at the US Gulf Coast and therefore had no immediate need to produce alumina locally. Based at Belmont, St. Ann parish, Reynolds mined, dried, and shipped both trihydrate and monohydrate ore from Ocho Rios Bay to their processing plants in the USA. Trihydrate (or gibbsitic) ore was sent to Reynold's alumina plant at Hurricane Creek, Arkansas, whereas the monohydrate (or boehmitic) ore was mined separately and dispatched to their plant at La Quinta Corpus Christi, Texas. Mined ore was conveyed to Ocho Rios Bay by an aerial tramway that carried 330 one-ton buckets over the 6 miles down to the coast (Zans 1952).

Frederick Snare Corporation, New York, contractors for the project installed the model designed by Ropeway Ltd of London. The tramway could carry 125 tons per hour at a speed of 150 yards per minute. Ore was loaded onto the specially designed self-loading ore ship *Carl Schmedeman* which could carry 13,400 tons. The ship was fitted to carry fuel oil for the drying plant on its return trip. The total round trip lasted eight and a half days between Ocho Rios and the Alabama State Docks at Mobile. The ship was outfitted with a unique conveyor belt system used for unloading that took about eight hours, one third the time for conventional bauxite ships. The ore mined between June and December 1952 when Reynolds started operation amounted to 239,948 tons after drying. This drying process removed up to 80% of the moisture, and after which it was then sent to Hurricane Creek. The first full shipment left on June 5, 1952. The ore, not as rich as that in Manchester, had approximately 49% alumina. Of the 50,000 acres of land held in St. Ann parish, only a tenth was bauxite-bearing, and the rest was dedicated to their large agricultural schemes. The company officially opened its head office at Belmont, St. Ann on January 9, 1953 (Zans 1952).

Early in 1950 Permanente changed its name to Kaiser Aluminum and Chemical Corporation and in 1952 the subsidiary Kaiser Bauxite Company was established. Construction of its facilities started in December 1951 and was completed towards the end of 1952. Mining began early in 1953. The company invested approximately US $12 million in their projects which was expected to last for 50 years. Like Reynolds the company dried and shipped ore to Kaiser Aluminum's plant at Baton Rouge, Louisiana which was modified and expanded to treat Jamaican ore. Up to 1953, 19,000 acres of land hosting the reserves to last for the 50 years were acquired in St. Elizabeth, Manchester, and St. Ann. The acquisitions also considered proposed expansions of the operations. Mining started at New Buildings, St. Elizabeth, in a large bauxite deposit between the May Day and Santa Cruz Mountains. This accumulation of trihydrate bauxite was 10 miles long, 2–3 miles wide and reached depths of 107 ft., though on average it was 10–15 ft. deep (Zans 1952).

Sixty-four rail cars, each carrying 70 tons travelled each day 13 miles from the mine to the ore handling, drying and storage facility at the newly constructed Port Kaiser at Little Pedro, St. Elizabeth. This amounted to nearly 4,500 tons of ore each day. The ore was placed on ships chartered from Marine Transport Lines Inc., New York, on destination to Baton Rouge where it was processed to alumina to supply the smelters at Chalmette, Louisiana, and Mead and Tacoma, Washington (Zans, 1952). Kaiser's first shipment of ore, 10,000 tons having 50.5% alumina, was loaded on the SS *Evanthia* in less than two days and departed Jamaica on February 9, 1953. It is said that persons from all over came to see the largest vessel that had ever docked on that side of the island. To memorialize the occasion the captain of the ship sounded the horn to let the onlookers know that the shipment was loaded and ready for departure (Neita 2009).

About 1958, Kaiser shifted its operations to Discovery Bay, St. Ann. As part of a partnership arrangement with two other companies, Kaiser contributed the facilities in St. Elizabeth. The tourism interests had strongly opposed the shift, but the residents welcomed the possible employment and economic opportunities. The Chief Minister Norman Manley was invited to intervene in the dispute. Manley, using a market stall as a stage, explained the benefits of having the bauxite industry. He concluded by asking the people if they wanted the industry to which he received a resoundingly positive response. Kaiser made its transition to Discovery Bay and established Port Rhoades, named after the company's vice president D. A. Rhoades (Neita 2009).

Towards the end of 1951 the British Aluminium Company Ltd had expressed interest in Jamaican bauxite. Prospecting work focused on those areas not staked by the other mining companies. Reconnaissance was undertaken in St. Catherine, Manchester, and Trelawny. The company also aspired to process some ore locally and engage in the exporting alumina and bauxite. They obtained their exclusive prospecting licence in May 1952 for 30 square miles in St. Catherine and conducted exploratory drilling. However, no high-grade deposits of economic quantities were discovered. The larger and more accessible areas with the better grades were already claimed and so British Aluminium abandoned its concessions in May 1953 (Zans 1952).

Bauxite mining and alumina production in Jamaica immediately ramped up and reached a peak by the mid-1970s. In 1952 when the industry started, 346,000 tons of ore was mined which represented 2.7% of the world's production. However, by 1958 Jamaica overtook Suriname as the world's number one producer with mine production reaching 5,184,000 tons or 27.2% of the world's production. By 1969, Jamaica was producing over 10 million tons, and this level would be maintained up to 1981, after which production fell dramatically (Roberts 1971; Government of Jamaica 1960; Davis 1995a; Mines and Geology Division Production Database). Davis (1995a) credits this production boom to the prevailing favourable economic conditions of the 1960s. The impressive growth rates in the developed world, low inflation, low oil prices, the policies adopted by the USA to support the Vietnam War as well as their domestic needs, along with general growing demand for aluminium including the penetration of the aluminium industry into other markets all contributed to the growth of Jamaica's Mining Sector.

In applying the guidelines suggested in the Memorandum on Colonial Mining Policies of the 1940s (Anonymous 1946), the Government had adopted a basic strategy for the Mining Sector, where Jamaica would be transformed from a mere supplier of raw materials into a provider of finished products. The main objective of this policy was to maximize income and the contribution to employment by the industry. As such, the Government was always concerned in obtaining a more equitable ratio between the bauxite shipped and that processed locally to alumina. In the early 1970s that policy began meeting success as the ratio trended positively. From 1952 to 1959, 17% of bauxite mined was converted to alumina. This increased to 23.5% from 1960 up to 1969. In 1970, when 12.1 million tons of bauxite was mined, 36% or about 4.3 million tons of bauxite were converted to alumina. It was projected that by 1975, 20 million tons would be mined, and that ore processed locally would be equal to or more that that exported. This was based on the firm plans agreed with the companies. At the end of 1970 the installed alumina capacity was 1.95 million ton per annum and the plans stated expansion to 2.9 million tons in 1973, and then to 4.35 million tons per annum by 1975. All the mining agreements concluded in the late 1960s between the Government and the mining concessionaires involved the production of alumina. One agreement limited the tonnage of ore that may be shipped, and another stated a provision for the supply of alumina to a smelter which would be constructed as soon as technical problems such as a source of cheap power are solved. The Government had prioritized having an aluminium reduction plant constructed in Jamaica (Lightbourne 1970; Walker 1971).

In 1974, all the bauxite companies exceeded their assigned minimum production. Bauxite production was 10% more than 1973 and peaked to an all-time high of 15.166 million tons (or 18.1% of the world's total production) representing 93.75% of the combined mining potential of all the local companies (Mines and Geology Division 1975; Davis 1995a). In the 70 years of bauxite mining in Jamaica, that is, up to December 2022 the optimistic projection of 20 million tons was never accomplished. Production in 2006 came close to the 1974 peak at 14.865 million tons but never surpassed that amount (Wacaster 2011).

By the end of 1970 there were six companies in Jamaica actively engaged in mining or setting up new facilities, namely Alcan, Reynolds, Kaiser, Alcoa, and the better-known copper producers Anaconda and Revere. The first four were among the largest in the world. Pechiney of France and Alusuisse of Switzerland did not come to Jamaica. Kaiser Bauxite Company was still involved in exporting ore from Discovery Bay and Reynolds Jamaica Mines Ltd did the same from Ocho Rios (Roberts 1971). Copper producer Revere entered the aluminium industry by establishing Revere Jamaica Alumina Ltd at Maggoty, St. Elizabeth. The company built an alumina plant and began exporting early in 1971 (Barker 1971). Kaiser, Reynolds, and Anaconda established the first joint venture in alumina production between American metal producers. This company, Alumina Partners of Jamaica mined bauxite and refined it to alumina at its plant at Nain, St. Elizabeth. The alumina was then exported from Port Kaiser at Little Pedro, on the south coast. When the Nain plant started in 1969 the initial plant capacity of 865,000 tonnes per year was the largest constructed in the world at that time (Barker 1971).

In late 1971 there were five alumina plants in Jamaica with total capacity of more than 2.8 million tonnes per year (t/yr). Alcoa was hesitant to enter the Jamaican mining industry when the other large companies were staking claims, probably to keep some distance from Aluminium Ltd. Aluminium Ltd took up concessions in Guyana and Jamaica. Alcoa on the other hand pursued deposits in Suriname and the Dominican Republic. Aluminium Ltd had evaluated the deposits in Dominican Republic but considered them uneconomic based on unfavourable terrain and inaccessibility, and shortage of resources such as water and labour needed for mining. Reynolds was given opportunity to develop the Surinamese deposits but opted for those in Haiti. Reynolds probably considered that the Haitian deposits, like those in Jamaica, were closer the USA, and the material may also have similar processing requirements and compatibility with their plants (Davis 1989).

Back in the mid-1950s Caribex Ltd, a subsidiary of American Metal Climax, Inc., was permitted to explore bauxite in the Mocho Mountains, in northern Clarendon parish, and by August 1957 they acquired concessions to mine in over 50 square miles. Caribex entered a joint venture with Alcoa which allowed Alcoa, through its subsidiary, Alcoa Minerals of Jamaica, Inc., established in July 1959, to exercise the option to take over the mining lease from Caribex by the end of May 1960. Alcoa exercised the option, and by 1960 the company was ready to begin installation of the necessary facilities for mining and exporting ore. Commercial production was expected to start by June 1963. The facilities constructed included an 18.8-mile-long railroad between the Mocho Mountains and a new deep-water port facility at Rocky Point near Salt River, Clarendon. Construction of this port involved the movement of 1.4 million cubic yards of material to reclaim 41 acres of swampland. The new dock also had a drying plant and storage area. A 3.5-mile-long causeway connected the port with the mainland. The project was completed several months late at a cost of US $14.5 million (Manley 1962; Davis 1989).

Alcoa, in December 1961, signed a 25-year agreement with the Government to mine bauxite from the Mocho Mountains (Manley 1962). Prime ore-bearing lands were already staked by the other companies and the Mocho Mountain was deemed 'very difficult country' (Manley 1962). Therefore, Alcoa, through seven Special Mining Leases, was permitted to stake reserves across six parishes. This included the Cockpit Country, which the other companies avoided due to the rugged topography and 'tight' cockpits. In total 288 square miles were secured. The first shipment of bauxite was exported in November 1963 (Davis 1989; Government of Jamaica 1963).

In 1968 Alcoa proudly announced plans to construct their new 500,000 tonne per year alumina plant at Halse Hall, Clarendon, and gradually end the export of bauxite as the plant became fully operational. Bauxite with less than 3% silica was prescribed as that which could be economically processed to alumina in the existing Jamaican and American plants. Alcoa, however, indicated that their plant would process ore averaging at 5.3% silica, ore previously classified as 'non-commercial' and representing a breakthrough for Jamaica's alumina processing. In addition, the agreement with Alcoa had a limit on the quantity of ore that could be exported. Alcoa, however, did not really have a choice of material as the better grades of ore were already staked. The alumina plant was commissioned in early 1972, a delayed start-up caused by work stoppage due to labour problems between the company and trade unions (Lightbourne 1970; Davis 1989).

Alumina Partners of Jamaica or Alpart represented the consortium formed in 1966 by Kaiser Corporation, Reynolds Jamaica Alumina Ltd, and Anaconda Jamaica, Inc., a subsidiary of the copper giant Anaconda. The consortium sought to utilize the reserves already staked by Kaiser and Reynolds to make alumina. By June 1966 plant design was underway by designers from Kaiser Aluminum and Kaiser Engineers. This plant was to be the most advanced in the world, incorporating innovative technology adopted from the other plants owned by the parent companies. Construction began in December 1966 on 300 acres at Nain, St. Elizabeth. This ambitious project cost the companies J$200 million. Additionally, Kaiser contributed the facilities and equipment at Port Kaiser, and bauxite lands, mining equipment and transportation infrastructure in St. Elizabeth and Manchester. Reynolds also contributed some of their bauxite lands it held in these parishes. On May 18, 1969, at 5:00 p.m. operations began and by August 9, 1969, the first alumina was shipped. The company initially struggled with problems related to mineralogical and chemical composition, particle size, and the presence of pisolites in the ore. However, by 1970 it was reported that the company had successfully overcome these hurdles. In addition to having the largest starting capacity in the world (that is, 865,000 tonnes per year), the Alpart plant was the most technologically advanced of any alumina plant globally. Alpart also had installed a 50 MW central power plant used for heating the huge digestion and evaporation vessels (Brown and Reckord 1970; Lightbourne 1970).

Anaconda had started in 1882 when Marcus Daly took over the copper mine at Butte Hill, Montana. The mine aided his fortune, and he became the first of the copper kings and a multimillionaire. The company diversified and by the 1950s it was the largest producer of zinc and copper. It also started the Jackpile mine in New Mexico, which was the largest uranium deposit in America. The company ventured into aluminium production in 1955 where it operated a smelter at Columbia Falls, Montana, being supplied with alumina from other producers. Anaconda also had fabrication plants in Indiana, Kentucky, New Jersey, and Georgia.

In 1965 the company announced plans to move upstream the aluminium chain by building a refinery plant in Georgia to extract alumina from clay (Brown and Reckord 1970). Kaiser and Reynolds recognizing the potential competition seem to have convinced Anaconda instead to partner with them in Jamaica. This would guarantee Anaconda a reliable alumina supply while protecting the interests and markets for Kaiser and Reynolds. Georgia's kaolinite clay reserves ran in the billions of tonnes and were being promoted for use in the aluminium industry, as it contains about 35% alumina. By using acid, the alumina could be removed from the clay, and the silica would remain insoluble. However, the Bayer process was well established, and it was more economical to obtain aluminium from the traditional bauxite which had more aluminium than kaolinite. By 1974, however, various companies began reporting their research into alternative materials. In February, Pechiney-Ugine Kuhlmann of France revealed a 20 ton per day pilot operation using a local 'non-bauxite material'. A month later Reynolds Metals Company conducted pilot studies at the Hurricane Creek, Arkansas plant using another non-bauxite material. In April, Alcoa and Anaconda disclosed an agreement to exchange information on work conducted on non-bauxite sources of aluminium. Georgia, in encouraging the use of its local clay for aluminium production, was willing to incentivize the mining companies. The research into the alternative sources of alumina may have been driven by news of upcoming changes in the terms of operations in the various developing countries where the companies had their ventures. Many countries began imposing policies and fiscal regimes that demanded more earnings being retained as was the case in Jamaica with the new bauxite production levy (Davis 1974).

4.4.1 The Struggles of Revere

Like Anaconda, Revere Copper and Brass also sought a share of the Jamaican bauxite industry. The fortunes, however, were not the same. Based in New York, Revere Copper and Brass was formed in 1928 in a merger of six small copper and brass companies. One of these, the Taunton-New Bedford Copper Company had acquired the assets of Paul Revere and Sons in 1801 and adopted the name. Revere focused on industrial fabrication of pipes, bars, tubes, and sheets. It had become one of the largest copper manufacturers in the USA. One of the original six companies in the 1928 merger was the Rome Manufacturing Company known for its tea kettles and houseware since 1892. Revere, in acquiring this experience, invented a famous line of cookware in the 1930s as it diversified from industrial applications. It developed a new, innovative design for its stainless steel pots. The new pots had a copper base for improved heat conductivity and faster heating. In addition, the handles were made of plastic; the first in the industry. The household items were marketed as Revere Ware, and Revere's housewares division was the most profitable (Anonymous 1998).

In 1966 Revere diversified into aluminium. The copper and brass industries were fluctuating rapidly but aluminium markets were more stable and profitable. Revere acquired 33% stake in Ormet Corporation, the primary aluminium producer and subsidiary of Alusuisse. Strategically the company undertook two projects, the construction of an alumina refinery at Maggotty, St. Elizabeth in Western Jamaica, to supply a new smelter and rolling mill at Scottsboro, Alabama. Maggotty was favourable because a rail already existed from Maggotty town through Jacobs Hut to Alcoa's Rocky Point port which the company arranged to utilize. The projects were small when compared with those owned by Alcoa, Reynolds, and Kaiser. However, Revere had plans for expansion (Anonymous 1998; Davis 1989). Unfortunately, the only experience they had in the aluminium industry was from the joint ownership of the Burnside, Louisiana alumina plant.

The subsidiary, Revere Jamaica Alumina Ltd acquired an exploration concession in St. Elizabeth and Trelawny held by Marvin Weaver Group of Texas. On March 10, 1967, an agreement was signed with the Jamaican Government where Revere would mine bauxite and construct an alumina plant capable of producing about 200,000 tons of alumina per year to be completed in three years, with capacity expansion to 600,000 tons per year within ten years. The company was later granted the 25-year Special Mining Lease no. 60 (SML 60) on April 10, 1967, to enable it to produce in total 20 million tons of alumina. This SML 60 was in the Cockpit Country. The alumina plant, constructed at Maggotty, St. Elizabeth, was expected to come online in 1970 but was commissioned in June 1971 due to problems encountered during construction. By the end of December 1974, US $97.6 million was invested in plant facilities and other fixed assets, which made it a grossly expensive venture. Additionally, designs were crafted by consultants who were unfamiliar with the challenges associated with treating Jamaican bauxite, rather than by experts from companies like Alcoa, Alcan or Pechiney. The Scottsboro smelter would be partially supplied with alumina from Maggotty, and it too was designed for expansion. In preparation, Revere entered an energy contract with the Tennessee Valley Authority which would facilitate the plant to triple in size (Lightbourne 1970; Davis 1989).

The Maggotty plant was inadequately designed and uneconomically scaled for profitability. The plant faced problems in the hydrometallurgical treatment of the bauxite, and this greatly reduced plant efficiency (Mines and Geology Division 1975; Davis 1989). The other bauxite companies mined haematitic bauxite, and from the 1940s had overcome their main processing challenges. The bauxite mined in St. Elizabeth in Revere's SML 60 was high in the iron mineral, goethite. This mineral retains a significant amount of alumina and requires high temperatures in the Bayer process, which results in higher energy costs. There were also settling problems. The plant could handle coarse-grained goethite, but fine-grained goethite did not react with the flocculants used to encourage settling. This fine-grained material was therefore not easily removed in the red mud waste stream (Zans 1952; Mines and Geology Division 1975).

The 1973 oil crisis compounded the problems associated with the cost of energy at Maggotty. Probably the least of Revere's problems were the complaints from the Maggotty community (Davis 1989). Between 1973 and 1974 fourteen complaints were filed against the company. The grievances included oil and alumina pollution to the Maggotty River and Black River, in addition to a pollution accident involving caustic red mud. The Mines and Geology Division, the newly formed Government mining regulator, found that approximately 125,000 gallons of red mud were discharged in the Black River due to malfunctioning equipment that caused an overflow of the mud washing tank. Other problems included citizens' complaints of dust from the haul trucks traversing through Maggotty town.

Between 1966 and 1976 Revere expended US $240 million on Maggotty and Scottsboro. In 1966, the parent Revere company's overall returns amounted to US $22 million from sales of US $350 million. By 1976 sales had increased to US $500 million but profits dwindled to US $2.2 million. Some problems were also external, such as the 1971 fall in copper prices and strikes at coal and copper mines that affected productivity. Revere declared a loss in 1972 as it wrote off surmounting debts from Maggotty. In 1973 the company reduced its copper production as it gradually lost its usual but vital supply of copper scrap. At the time, copper scrap sold on the US markets for US $0.60 to US $0.68 per pound but fetched over US $0.95 per pound on the London Metal Exchange. Scrap dealers therefore gravitated to preferential selling on the international market resulting in Revere Copper and Brass struggling to acquire scrap to remain profitable. This was despite the prevailing high demand for copper products (Anonymous 1998).

Probably the greatest problem faced by the struggling company was the introduction of the 1974 bauxite production levy in Jamaica. By 1974, Jamaica was suffering from high inflation, and it was necessary to increase revenues. The Government decided to introduce a tax on bauxite mined. For weeks, secret negotiations were held with the companies to determine an amicable rate. However, the Government brought the matter to conclusion in March when it served each company with the specific terms for revenue and local participation. For Revere this included a production levy effective and due as of January 1, 1974. This amounted to J$14 per long dry ton of bauxite mined and processed into alumina in Jamaica. Royalties were replaced with a flat rate of J$1 per long dry ton of bauxite processed locally. Depletion allowances on the bauxite reserves were discontinued effective January 1, 1974. Premiums to the Overseas Private Insurance Corporation (OPIC, an insurance firm protecting American businesses from failure due to Government actions such as nationalization) would not be allowed as a deductible expense. In addition, it was stated that the Government 'reserves the right' to disregard contractual provisions calling for most favoured company treatment. Further, all foreign exchange 'must be sold' to an authorized depository and all overseas payments 'must be made' from a Jamaican bank account and subject to Exchange Control approval (AAA 1978; Anonymous 1998).

Revere had managed to conclude a deal in May 1974 with a consortium of six Japanese companies where US $160 million would be injected into Maggotty to double production towards making it economical to operate. When informed of the levy the consortium withdrew, and Revere was stranded. Revere reluctantly paid the tax hoping that the Government would grant some concessions. For the other companies, they were already established (and quite profitable) and as such were able to operate despite the tax (Cross 2013). The concessions never came, and Maggotty was shut down in August 1975. On August 19, 1975, the company issued a statement that it was suspending operations at Maggotty as a 'result of curtailment in primary aluminum production in the USA'. The suspension, which it deemed temporary, was effective immediately and 550 Jamaican workers were laid off. Revere indicated that resumption of the operations depended primarily on the timing of economic recovery in the USA and improvement in aluminium demand (Anonymous 1975).

Shutdown of the operations was not a simple task as the agreement with the Government required Revere to obtain permission for closure. Additionally, some taxes were still due. Revere offered to sell Maggotty to the Government for US $65 million, but this offer was refused. In January 1976 the company stopped paying taxes and sought intervention of the Jamaican Court to be freed from the debts. Revere also made a claim in April with the OPIC where it defended that the Jamaican Government should seize the plant for the levy owed. It lost the case in the Jamaican Court. However, as a consolation in 1978 Revere was awarded some compensation from the OPIC based on the results of the matter being heard by the American Arbitration Association (AAA 1978; Anonymous 1998).

Maggotty was proving even more burdensome as it could not be otherwise sold. The company soon offered up Scottsboro for sale, and it was gradually appearing that Revere would be sold overall. Alcan offered to buy Scottsboro for US $205 million in July 1977.

The sale was stopped by the US courts for antitrust reasons, and in December 1977 the agreement was cancelled. Revere was stranded yet again. In the 1980s Revere filed for bankruptcy under Chapter 11 of the Federal Law citing 'massive and continuing losses'. The company managed to sell some of its aluminium fabrication facilities and retained only the housewares division which was profitable. Through the law the company could continue to operate while making attempts at debt repayment. This provision was sought in November 1982 where just months earlier Scottsboro was eventually shut down.

By 1983 one-third shareholder in Revere, Asarco, a giant metals conglomerate sold its stocks to Bear Stearns and Company, of New York for US $17 million. Soon other investment companies bought on Revere stocks, improving on its housewares division through marketing and new innovations. By April 1988 Revere Copper and Brass divested its housewares division to popular glass cookware manufacturer Corning Glass which paid US $120 million for Revere Ware. Corning was well known for brands like Corning Ware and Pyrex (Anonymous 1998). The Maggotty plant remained a relic on the Jamaican landscape until it was gradually scrapped of its equipment. In the 1990s the Maggotty site was used by a company that mined limestone and produced quick lime to supply the JAMALCO alumina plant. However, in June 2003 this enterprise was placed in receivership by its funders, the National Investment Bank of Jamaica, for failure to repay its debts (Thompson 2017).

4.5 Bauxite Production Levy

Since the start of commercial bauxite mining and alumina production in Jamaica in the 1950s, that industry had made significant contributions to the country's economic growth. From 1952 up to the end of 1970, approximately 116,270,000 long tons were mined. Of that amount 27.5 million tons were converted to 11.3 million tons of alumina. In total the country earned J$1.1 billion where J$504 million was for the bauxite exported, and J$596 million for the alumina shipped. The Government collected a total royalty of J$25.3 million from these exports, not considering the personal and corporate income taxes and other contributions to the Government and the economy which exceeded this figure. The Bauxite and Alumina Industry had contributed approximately 16% to the country's Gross Domestic Product in 1970 and was viewed as the economic sector with the greatest potential for growth. It was foreseen that by 1973, with the planned expansions, the share would reach 20% (Barker 1971; Tretzel 1971; Hill 1971; Walker 1971). However, the companies, over the years, were able to minimize their tax burdens by employing creative accounting techniques such as transfer pricing. It was reported that Alcan paid as little as one-seventh what it owed the Jamaican Government in taxes and duties. Up to the 1970s this would have amounted to annual revenues of approximately US $70 million (Cross 2013).

In addition to the country being shortchanged, Jamaica faced a loss of competitiveness of the bauxite industry, which traditionally was a driver of economic growth, being a main source of both revenue and foreign exchange. This was coupled with adverse balance of payments resulting largely from the OPEC-induced increases in oil prices in the early 1970s (Clarke 1979). Immediately after his election in 1972, Prime Minister Michael Manley initiated a review of the bauxite and alumina industry. In October 1972, he told the United Nations General Assembly that

> The time has come to reconstruct the basis on which the gains arising from the exploitation of . . . resources are shared between those . . . who provide the capital and technical know-how for the process of exploitation . . . and those . . . who own the resources and provide essential infrastructure and the labour force. (AAA 1978)

He later stated that throughout 1973 Jamaica was setting the stage at home and abroad for the actions to be taken in 1974. He had established a National Bauxite Commission in 1972 and conducted 'an intensive investigation of the aluminium industry' in 1973 which led to the development of a policy with four objectives. These included: (i) a drastic increase in revenues from bauxite mining and alumina production; (ii) recovery of bauxite ore leased to the mining companies; (iii) reacquisition of all lands owned by such companies; and (iv) national majority ownership and control of the bauxite industry. The Government also had plans to strengthen its institutional capacity for better management of its interest in the industry.

As such, on the morning of March 15, 1974, the Prime Minister announced the sudden and unilateral declaration of a new tax called the bauxite production levy, the repatriation of bauxite lands, and the strictest bauxite operating rules. The reformed fiscal regime substantially increased revenues from the industry, but also made the cost of Jamaican ore greater. The production levy was made effective and payable from January 1, 1974 and was calculated using a formula using the average of the price realized by Alcoa, Reynolds and Kaiser for one short ton of primary aluminium (as defined in their Form 10-K reports to the US Securities and Exchange Commission; this document gives a comprehensive summary of the company's financial performance) and the conversion factor of the amount of bauxite used to make a short ton of aluminium (Bell 1979; Clarke 1979; Davis 1995b; Barclay and Girvan 2013; Das and Patel 2013).

It is reported that investor trust and confidence in the Government collapsed. The bauxite companies retaliated to this policy shift. They complained that these taxes would make their operations uneconomical and threatened to leave. Three of the six operating companies sought dispute resolution at the World Bank's International Centre for Settlement of Investment Dispute. Production fell abruptly from the 1974 high of over 15 million tonnes, to 11.379 million tonnes, and the companies also began restricting investments in Jamaica which resulted in diminishing plant capabilities (Mines and Geology Division Production Database). The Revere alumina plant having 200,000 tonnes capacity was closed in August 1975. Alpart started producing below rated capacity and cited the need for a major capital injection to produce at near capacity levels. Alcan, in 1976, started operating only three of their four units. Kaiser began purchasing some 500,000 tonnes of bauxite from another country. And finally, Reynolds scaled back production from 3.7 million tonnes in 1974 to between 2.5 and 2.8 million tonnes per annum. In addition, they accelerated their earlier strategy to expand investments in Guinea or Australia. The plan also involved divesting themselves of the Jamaican operations (Clarke 1979; Barclay and Girvan 2013; Neita 2022).

Despite citing the levy, the transnational companies were not truly planning to invest in any future expansion in capacity and production in Jamaica. The companies had developed the strategy to locate incremental production in one location and subsequently expand production in new locations. The North American companies located incremental production in Suriname and Guyana in the 1920s to late 1940s, then in Jamaica and to a lesser extent in the Dominican Republic and Haiti in the 1950s and 1960s. The companies then concentrated on expanding capacity in Australia and Guinea from the 1970s. Reynolds shuttered its mining operations in Jamaica in late 1984 and withdrew from the Jamaica Reynolds Bauxite Partners joint venture. The company cited that the protracted decline in the international aluminium industry forced it to take advantage of the alternate resources in Australia, Brazil, and Guinea. The Jamaican Government therefore acquired all the assets (Barclay and Girvan 2013).

Alcan, in 1971, had lost its bauxite interests in Guyana to nationalization. That company had already begun shifting its primary mining operations to Jamaica during the late 1960s, a role it was preparing Jamaica for since the 1940s. The company was now contending with the disclosure of under-reporting and underpayment of the duties on their exports. This was in addition to avoiding their large tax bill for their mining operations. The income denied to Jamaica was calculated at equal to the budget shortfalls that Jamaica was experiencing in the mid-1970s. The company recognized that the fiscal reform in

Jamaica was less threatening than nationalization and so continued to operate while preparing the reserves in Australia to become the primary operation (Cross 2013).

The decline in production in Jamaica though largely attributed to the imposition of the levy, was also driven by some other economic, technical, and social factors.. These included an economic recession in 1975 to 1978 and the significantly low global demand for commodities like aluminium. There was the widespread availability of bauxite in Australia, Guinea, Brazil, and India which provided the companies with options to diversify supply, and these other countries had fiscal regimes that were more competitive. The companies had also developed the technical ability to substitute bauxite varieties in their plants. Further, Jamaica did not have indigenous energy resources when compared to the major competitor like Australia which had local coal that fuelled their largest alumina plant. Locally, there was industrial unrest in 1975 and 1976 which also impacted the industry (Clarke 1979).

Despite the protests, the companies recognized that the levy and other policies would definitely be imposed. They also recognized the advantageous factors such as geographic location that Jamaica offered when compared with the alternatives (Clarke 1979). In mid-1974 Kaiser's Board President Cornell Maier had declared that

> What we seek in Jamaica is not land, it is bauxite. We believe that ownership of bauxite lands can be good for Jamaica and for the company. We agreed, as we had agreed from the beginning, that we wanted to find a way to agree that the best interests of Jamaica and our company could be served no other way. (Neita 2022)

And

> What we have sought in Jamaica in our long history here is bauxite... and the friendship and respect of the Jamaican people. (Neita 2022)

Participation agreements were therefore made with Alcoa in 1976, Kaiser in 1977, and Alcan in 1978. The companies also called for a review of the fiscal regime citing the effect of the levy on the total cost of their operations and the competitiveness of their product. After 18 months of discussion and negotiations a new levy formula was agreed and became effective in July 1979. The new levy utilized a production incentive formula which was so designed to maintain a desirable inflow of revenue (as a safeguard to prevent reductions in production) and to restore the idle capacities (Clarke 1979). The levy regime was subsequently revised in 1984 and 1988. A later revised regime which currently obtains, employs the price of primary aluminium on the London Metal Exchange (LME). Since the Second World War the aluminium transnational corporations had considerable control over the pricing of aluminium. However, in the 1970s a limited free market in aluminium developed, and the influential transactions attracted the attention of the LME resulting in it establishing a contract in primary aluminium ingots in 1978. This introduction of the LME for aluminium destabilized the prices in the industry and the prices began to exhibit cyclicality like other metals such as copper. The aluminium transnationals had reduced influence over the stock levels and were unable to significantly affect price levels (Barclay and Girvan 2013).

Prior to the imposition of the bauxite levy in 1974, Jamaica had boomed and peaked in bauxite production. However, the contribution to total Government revenues by the entire industry had reached a measly maximum of 8% in 1973. When the levy was imposed, the total contribution rose to some 37% of total Government revenues. The bauxite production levy led to a revenue increase of approximately US $10 more per ton, but the price of bauxite increased from US $8 to $20 per ton. The levy, therefore, at that time resulted in a five- to six-fold increase in Government revenues relative to the average returns of 1960–1973. Between 1974 and 1978, US $845.2 million was collected, with over US $250 million collected in 1978 alone. This remarkable financial accomplishment in the bauxite industry was memorialized on the country's ten-dollar bank note (Fig. 4.1). The funds were placed in a special account called the Capital Development Fund for the purpose of investing in economically

Fig. 4.1 Reverse of the Jamaican $10 Bank Note first issued in 1978 showcasing bauxite mining. During this time the Government had increased state participation in the bauxite industry. The nest of precipitator tanks at the Kirkvine Works at Shooters Hill, near Mandeville, Manchester is shown, with the Mile Gully Mountain in the background. This alumina plant was the first to be constructed in Jamaica (and the Caribbean), between 1951 and 1952. The various mechanical equipment used in mining and transportation are in action, such as dragline excavators with extended booms (one partly obscured to the right), 'Caterpillar' ore-carrier wagons (*right*) and bulldozer (*left*) (Reproduced with permission from the Bank of Jamaica)

self-sustaining ventures. However, this fund was regularly drawn down to support the national budget as was the case during the same period of 1974 to 1978 where US $538.6 million was transferred to the Consolidated Fund to finance the country's expenditures. Since the imposition of the levy up to 1988 US $2.158 billion was collected and 65.8% was drawn down to finance the budget (Bell 1979; Clarke 1979; Davis 1995b; Barclay and Girvan 2013; Das and Patel 2013). The Government continues to use this fund for budgetary support, with annual withdrawals made to sustain the operations of a Government special purpose vehicle for mining, and to finance development projects in mining communities (Angus 2020; Linton 2023). Occasionally, a large-scale draw-down may be necessary, such as in 2004 when J$667.5 million was withdrawn to supplement the Consolidated Fund (Government of Jamaica 2004). As at March 31, 2023 the Capital Development Fund had J$4.9 billion (Linton 2023).

4.6 State-Owned Enterprises

Following the imposition of the levy, the Government conducted negotiations with the companies with the intention of state participation in their operations. These agreements were concluded between 1976 and 1978. The Government had recognized that the operational involvement required a range of special purpose vehicles or public companies dedicated to protecting the Government's interest in the bauxite industry. This resulted in the establishment in the mid-1970s of the Jamaica Bauxite Institute (JBI), the Jamaica Bauxite Mining (JBM), and the Bauxite and Alumina Trading Company (BATCO).

The JBI was established with the mandate to monitor the bauxite industry on behalf of the Government and conduct economic and technical research towards its development. Jamaica was also considered to be at a disadvantage when negotiating with the bauxite companies.

The experience of the previous negotiations revealed a scarcity of relevant data, a lack of analytical capacity, and a general unpreparedness when bargaining with the transnational companies. At the 1953 negotiations, the Government failed to get an increase in royalty from US $0.14 to US $0.28 per ton of bauxite, and this, it is said, was due to the lack of data to support the proposals. Fortunately, however, at the 1956/1957 negotiations more information was available as the Government had hired international consultant, Samuel Moment. The successes resulted in Moment serving as advisor on bauxite to Jamaica between 1956 and 1974. The need for a dedicated organization became even more pronounced as the Government, with the policy shifts in 1974, negotiated significant increases in income and was about to actively participate in the operations of the companies (Davis 1995b; Barclay and Girvan 2013).

It was proposed that a bauxite institute with centralized bauxite functions supported by natural scientists, engineers, and economists as the core staff, should be established. This would be a research and development organization to deal with all non-regulatory bauxite matters (regulating the industry resided with the Mines and Geology Division). This included conducting research to optimize the consumption of Jamaican bauxite, keeping abreast of developments, including the economics of the global aluminium industry, and providing technical support in negotiations. The entity would be a technical and economic knowledge base to support the work of the National Bauxite Commission (and the Bauxite Negotiating Team). In addition, the JBI would keep the tally of the amounts of bauxite production levy payable by the companies to the Capital Development Fund for reporting to the Ministry of Finance. Therefore, on October 27, 1975, the Cabinet decided that the JBI should be established as a subsidiary of the Jamaica National Investment Company. The Mineral Research and Development Division of the Mining Ministry was reorganized in 1976 to form the JBI. The Institute was first based at Cross Roads in Kingston, then relocated in 1982 to the Government office complex at Hope (Davis 1995b; Mines and Geology Division 1975).

With the relocation to Hope, the JBI was able to expand its practical research capabilities with the installation of several pilot facilities for material testing. These include a small Bayer process refinery installed in 1984, a rare earth element (REE) extraction facility installed in 2013 and a plant nursery for agricultural research. The bauxite companies utilize the pilot-scale Bayer plant to conduct processing trials of different varieties of bauxite, in addition to testing new reagents with small batches of bauxite before introducing it at their larger plant. In more recent times a REE plant was constructed to test the extractive metallurgy of rare earth elements from red mud waste (JBI 2021a).

In 1996, the Government moved to allocate a portion of the Capital Development Fund towards social investment in the bauxite communities to create and sustain 'life after bauxite'. As such, the Bauxite Community Development Programme (BCDP) was established to implement projects to improve the way of life in the bauxite communities. Part of this project involved investigation of various vegetable crops that can be successfully cultivated on rehabilitated mined-out lands. Greenhouses were erected at Hope to nurture seedlings and other plants to be used in farming rehabilitated bauxite lands. Farmers' cooperatives in the bauxite areas were equipped with greenhouse facilities on rehabilitated lands to operate and make productive. Some ore pits were also shaped to be utilized as rainwater harvesting catchments and storage ponds to supply the agricultural activities. The BCDP is administered by the JBI and since its inception, to 2022, seven phases have been implemented and nearly J$600 million has been expended towards the implementation of over 350 projects (JBI 2021b).

The JBI model of research at every stage of mining was adopted by the Indian Government to establish their Jawaharlal Nehru Aluminium Research Development and Design Centre in Nagpur in 1989. Support for this establishment came through partial grant funding of 170 million rupees from the United Nations

Development Programme. Like Jamaica, India too had learnt lessons from their experience from the 1950s to the 1970s, and also sought to ensure that the country had a strong position in managing its bauxite resources (Das and Patel 2013; JNARDDC 2018).

The National Bauxite Commission supported by the JBI represented a body of distinguished local specialists appointed to review the conditions of the industry and of the resources, and to advise the Mining Minister towards ensuring that the exploitation of the bauxite resources makes the maximum practical contribution to the long-term development of the country. It was established in 1972 with members from both the public and private sectors along with a representative from the University of the West Indies. Industrialist, Mayer Matalon, served as chairman, and his deputy was corporate lawyer, Patrick Rousseau. The commission's mandate involved assessing the bauxite resources, planning the use of the lands, securing training for Jamaicans to run the industry, and planning for and implementing the partnership with the bauxite companies. The commission was also responsible for implementing the proposals for smelter development and the establishment of fabricating plants for aluminium products (Davis 1995b).

In 1975 the Government established the JBM as the holding company for the Government's equity, and the representative in the partnerships with the bauxite companies. The agreements concluded between 1976 and 1978 provided that the Government would control 51% of the mining operations of the companies. This was straightforward for Reynolds and Kaisers which simply mined bauxite. For the alumina operations of Alcoa and Alcan, the 51% was converted into a 6% and 7% share respectively of their overall bauxite mining and alumina processing operations. The Government also repurchased the reserve lands held by the companies at a written-down book value. The payment for the land would be over ten years and in return the Government would guarantee the respective companies sufficient reserves to sustain operations at their existing capacities for a 40-year period. A mining lease would be provided for the reserves allocated (Clarke 1979).

The BATCO, established in 1977, utilized a team of specialists to trade bauxite, alumina, and oil for the Government. The company strategically marketed and traded bauxite and alumina for JBM and later, the Clarendon Alumina Production Ltd (CAP). In the 1970s new markets were established and sales agreements settled. Venezuela purchased one million tonnes of alumina, and other sales were made to Algeria, Iraq, and the USSR. There was also a counter-purchase agreement with the Ukraine where, between 1983 and 1991, Jamaica exported a small percentage of bauxite to the Ukraine and was able to purchase motor vehicles such as Lada motorcars and other goods from Russia, in addition to receiving much needed foreign exchange from bauxite sales. The expiration of the counter-purchase agreement resulted in the bauxite trading department being disbanded and trading limited to alumina and oil. Oil was traded on behalf of CAP (Barclay and Girvan 2013).

BATCO became the Government's vehicle for the forward sales of bauxite and alumina. Sometimes, forward sales contracts were entered as support for Jamaica's budgetary deficits especially during financial crises. Since about the late 1950s Jamaica had entered a Bauxite Barter Programme with the USA where large quantities of ore were bartered for US agricultural surplus commodities which are disposed of elsewhere in the world. Up to 1962, one million tons of bauxite were already sent to the stockpile programme, and the revenues obtained by the Government of Jamaica from the transaction amounted to about £700,000 per annum. The programme provided a dedicated market that sustained the mining operations and resulted in employment and commerce. The Jamaican Government therefore encouraged the US Government for its continuation. (Sangster 1962). Even in the 1980s Jamaican bauxite was supplied to the US stockpile. In 1981, through a forward purchase the US Government obtained 1.6 million tons of Jamaican bauxite. By 1984 several other purchases had resulted in a total of 11.45 million tons being acquired, mostly for the stockpile rather than for commercial production.

Another example is the forward sales to Swiss commodity trader Marc Rich & Company AG. This company began dealings with the Jamaican Government in the 1970s when it and another company called Derby alternated in managing the alumina sales of BATCO to the Government of Venezuela's company VENALUM. Marc Rich and Company AG later became Glencore. In the mid-1980s the Clarendon Alumina Works (JAMALCO alumina plant) which was operated by Alcoa had closed resulting in a sharp drop in income to the country. The Government established CAP in 1985 to operate the JAMALCO plant. The plant was reopened in 1986 under a three-year lease arrangement. The Government entered a 10-year agreement with Marc Rich, who purchased approximately 4.119 million tonnes of alumina at 25% below the market rate. Alcoa recognized that the JAMALCO operation being run solely by CAP had significantly improved and decided to return to Jamaica. During the years of managing the JAMALCO plant, the Jamaican Government was able to make it more efficient. Production costs decreased by almost 50% and significant profits were being made. In March 1988, Alcoa resumed production for itself, with the Government acquiring 44% of Alcoa's share to increase total ownership to 50%. In 2002, the Government again entered into a 10-year forward sales agreement for alumina with Glencore. This time the country received a US $65 million loan to finance CAP's investment in the expansion of JAMALCO to 1.25 million tonnes alumina per year. This agreement, like that before, priced alumina below the current market rates and the current production costs, and made no allowances for increases in energy and production costs, which rose considerably after the agreement was signed (Davis 1995b; Davis 2011; Barclay and Girvan 2013; Farchy 2017).

In February 2013, CAP concluded the longstanding agreement with BATCO and as a result, BATCO ceased to manage alumina supply agreements on behalf of CAP. BATCO would no longer benefit from the commission on CAP's alumina shipments from JAMALCO. This resulted in BATCO becoming obsolete, and in February 2014 the Government approved the winding up, the transfer of assets to JBM, and the forgiveness of a US $16 million debt owed by JBM to BATCO (Collinder 2014). The winding up started in mid-2014 and liquidation of the assets began until it stalled in 2017. The winding up process was restarted in 2022.

4.7 International Bauxite Association

Jamaica's bold policy shifts towards the fiscal regimes and strengthening of the institutional arrangements for the management of the bauxite industry was coupled with the country actively leading in the establishment of the International Bauxite Association (IBA). This producer association targeted increasing the economic rents payable to the bauxite-producing countries. In early March 1974 a high-level international conference of bauxite-producing countries convened at Conakry in Guinea. In attendance were ministers and other officials from Australia, Guinea, Guyana, Jamaica, Sierra Leone, Suriname, and Yugoslavia. Algeria, Cameroon, Ghana, and Mali sent observers. Indonesia was invited but did not attend. Brazil did not join the association. At the meeting the establishment of the IBA was ratified with signing of the agreement on March 8, 1974. The headquarters was in Jamaica. The IBA's objectives were to promote the orderly and rational development of the bauxite industry, to secure for member countries fair and reasonable returns, and to generally safeguard the interests of member countries. (Anonymous 1974). In 1973 the IBA member countries accounted for 69% of global bauxite production, and this reached 74% in 1974 (Barclay and Girvan 2013).

The IBA was established in a period of resource nationalism in developing countries and was modelled on the Organization of the Petroleum Exporting Countries (OPEC). It was intended to emulate the success in securing dramatic increases in the returns to producing countries from exports of natural resources by cartelizing the market and increasing bargaining power when negotiating with the transnational companies. However, the Association began to fail in its objectives. The IBA was not able to cartelize the bauxite market and had no significant impact on pricing and processing, nor could it control the transnational companies in their markets. Further, the member countries could not agree on a standard system of prices and taxes for bauxite and alumina exported from the members, nor could they maintain world market prices by means of global production limits (Barclay and Girvan 2013).

The member countries did not recognize that bauxite is not a homogenous commodity that is freely traded on world markets. The chemical and physical properties of the ore vary with origin and the processing plants are customized for treating unique ore properties. Production, processing, and transportation costs also vary widely based on the region. These technical challenges led to member countries struggling to agree on a formula for determining reference prices. Further, the members met contradictions in interests. Older producers such as Guyana and Jamaica wanted to maximize returns from existing production. Newer producers such as Australia and Guinea sought to maximize their returns through new investments, higher production, and increasing global market shares. The collapse of the IBA began in 1991 when Australia withdrew. Member countries lost interest, defaulted on their contributions and meeting attendance dwindled. The Association was formally dissolved in 1994 when Jamaica withdrew (Barclay and Girvan 2013).

4.8 Jamaica's Bauxite Industry in Recent Times

The Jamaican bauxite industry has evolved with the assets changing ownership as companies adopted various business models. Reynolds met its exit strategy timeline and pulled out from Lydford, St. Ann in December 1984. Back in 1969 the company was part of the Alpart joint venture with Anaconda and Kaiser. However, Anaconda divested, and Reynolds acquired its shares. Reynolds maintained partnership with Kaiser at Alpart until 1985 and obtained bauxite from Kaiser's Discovery Bay operations for its Gramercy and Corpus Christie refineries. Gramercy was closed for a while but the Jamaican mines continued supplying Corpus Christie (Davis 2000).

In 1999 Alcoa presented a bid to take over Reynolds' refining, smelting, and fabrication facilities which included shares held in companies in Guyana, Guinea, and Brazil, along with a purchase agreement for Australian bauxite up until 2021. However, a condition of the takeover required Alcoa to dispose of all of Reynolds' alumina assets. At that time this included the full ownership of the alumina plant at Corpus Christie, Texas, a 56% share in the Worsely, Australia alumina plant, and a 5% stake in the alumina plant at Stade, Germany. Alcoa had to also divest 25% of the smelter at Longview, Washington. The deal was completed in May 2000 at a cost of US $4.4 billion with Reynolds being incorporated into Alcoa which again became the largest aluminium company in the USA and one of the world's top ranking aluminium producers (Davis 2000).

Alcan's departure from Jamaica came in 2001. In May 2001, Glencore International acquired Alcan's 93% stake in the joint venture operations with the Jamaican Government. The parent company Alcan later became incorporated with the British-Australian multinational mining corporation Rio Tinto PLC in 2007 for US $38.1 billion. Alcoa had attempted a US $33 billion hostile takeover of Alcan in May 2007 stating its aim was to create a 'premier global aluminium company with enhanced growth opportunities'. Alcoa would have gained status as the world's largest aluminium producer. The bid was withdrawn when Alcan declared acceptance of Rio Tinto's friendly takeover bid in July that year. The takeover was completed by October 2007 (CBC 2007; Rio Tinto 2007).

Glencore renamed the Jamaican operations West Indies Alumina Company or WINDALCO and gained control of the two alumina refineries, two bauxite mining operations, and the shipping dock. Glencore had also acquired Kaiser's 65% stake in Alpart through a secondary sale from Norsk Hydro. It held these assets for several years until they became the subject of a merger. In October 2006 it was announced that Glencore's alumina assets in Jamaica, Ireland, and Italy, along with a smelter in Sweden were to be merged with the assets of Russian companies RUSAL and SUAL (RUSAL 2007). In April 2007 the merger was completed to form United Company RUSAL (UC RUSAL), the world's largest aluminium and alumina producer at the time.

With UC RUSAL's entry to Jamaica there were plans for increasing plant capacities. Alpart would be upgraded from 1.65 to 2 million t/yr. WINDALCO's production level would be increased above the 1.2 million t/yr (JIS 2007). The world-wide financial meltdown of 2008, however, affected the bauxite industry and by 2009 UC RUSAL began scaling down operations in Jamaica. The Ewarton refinery was the first to be shut down in March 2009. Low aluminium prices did not help, and soon Kirkvine was shut down in April, and Alpart in May. Ewarton, however, was restarted in June 2010 under terms that the company received a full waiver of the production levy which averaged at US $5 per tonne of bauxite or bauxite equivalent mined for the first year and between US $2.50 to $3 per tonne thereafter. UC RUSAL however reportedly sought additional financial concessions more than twice as good as those received for opening Ewarton, for reopening Kirkvine which was operating at 550,000 t/yr capacity. UC RUSAL sought a two-year waiver on the levy. The Government however offered the same terms as before, but Kirkvine remained closed. Yielding to UC RUSAL's demands would have made it more difficult for the Jamaican Government to meet fiscal deficit targets demanded under its loan agreement with the International Monetary Fund (IMF). In February 2010 the Jamaican Government was faced with significant reduction in revenues especially from the bauxite and alumina sector and so it entered a US $1.3 billion loan agreement with the IMF. The IMF wanted Jamaica to divest itself of its bauxite assets which were the origin of mounting debts (Anonymous 2011; Porter 2012a). The Jamaican Government in June 2014 sold its 7% share in WINDALCO to RUSAL for US $11 million. This amount was applied to a remaining US $21.15 million debt Jamaica had to RUSAL, under an alumina forward sale arrangement with Glencore that failed when alumina prices plunged. The remainder of the debt was settled by JBM (Reynolds-Baker 2014).

By July 2014 the Government threated UC RUSAL to revoke its mining leases for bauxite for both Kirkvine and Alpart if they failed to restart in six months. The Government further indicated that the lease for reserves supplying Ewarton would even be revoked if UC RUSAL did not make any attempts. UC RUSAL responded that the restart would 'not be possible without time-consuming and expensive re-equipment and modernization' necessary for converting the plants to coal or gas-fired (Luton 2014a). The reopening of Kirkvine was continuously stalled citing the need for an energy solution. Over the years the plant was mothballed and scrapped.

In September 2011 minority shareholder in the dormant Alpart plant, Norsk Hydro had sold UC RUSAL its 35% share for US $46 million which allowed the latter full control in the company and 65% of alumina capacity in Jamaica (Jiang 2011). UC RUSAL finally reopened the Alpart plant in early September 2015 after having been closed since May 2009 due to high production costs arising from the global recession. The operation, however, was limited to the exporting of raw bauxite, and the first shipment of 35,000 tonnes of ore was loaded at Port Kaiser. UC RUSAL sought to export two million tonnes of crude wet bauxite over an 18-month period to supply a refinery in Russia (Al Circle 2014; Davis 2015). By the end of 2016, however, UC RUSAL announced that in November it had completed the sale of its 100% stake in Alpart to the Chinese state industrial group Jiuquan Iron and Steel (Group) Co. Ltd (JISCO) for US $299 million (RUSAL 2016).

The year 2018 turned out to be challenging for UC RUSAL. In April, the US Department of the Treasury's Office of Foreign Assets Control (OFAC) announced its intention to apply sanctions to seven Russian oligarchs which included billionaire commodities magnate Oleg Deripaska, the single largest shareholder in UC RUSAL, which is the parent company of West Indies Alumina Company (WINDALCO) and En+Group PLC, a Jersey (UK) company holding 48.13% equity interest in UC RUSAL. As a result of being on the list, all assets of Deripaska and entities connected to him that were within US jurisdiction were frozen, and US persons were prohibited from business dealings with them. Non-US persons stood to face secondary sanctions for knowingly facilitating significant transactions for, or on behalf of, the individuals or entities sanctioned. On April 23 the USA issued General License 14 (GL-14) that provided for maintenance and winding down activities. This GL-14 expired on October 23, 2018.

The Jamaican Government, to reduce the effects of the sanctions on the more than 1,200 workers at WINDALCO, and to minimize the potential negative impact on the livelihood of the Jamaican citizens and the local economy, made an appeal to the US Department of the Treasury in June 2018. The Government sought a Special License in case the sanctions took effect after GL-14 expires. The OFAC responded that if the activities of WINDALCO are consistent with those authorized by GL-14, WINDALCO could continue to operate unaffected. It was reported that in May 2018 Deripaska had disposed of his shares and in June the En+board of directors had changed. The US Embassy in Jamaica, however, indicated that they were unaware of Deripaska's actions. By early 2019 the sanctions were lifted following a shareholder revamp that handed control of UC RUSAL to En+Group. Deripaska, however, sued the US Government for the sanctions which he claimed had cost him billions of dollars, made him 'radioactive' in international business circles, and negatively impacted his affairs in Russia. The US court ruled that Deripaska's numerous claims lacked merit (Gittleson 2019; Thomsen 2019; Vogel and Rappeport 2019).

Unfortunately, it was just a few days after the US Government announced the sanctions that would affect UC RUSAL that on April 11, 2018, the Mining Minister revoked the mining lease for the Kirkvine plant on the basis that the refinery was closed since 2009, and the company did no 'work' (mining) as required under the Mining Act. It was also reported to the Government that the Kirkvine plant was being scrapped. The company admitted to suspending the extraction of ore from Kirkvine from 2009 to the time of the revocation of the lease, but this was due to circumstances such as market fluctuations and poor plant energy efficiency which were beyond their control (Luton 2014a; Wilson and Barrett 2018).

UC RUSAL brought the Government to court asserting that at no time did it cease work on the land for an extended period as there was work ongoing in maintaining the plant equipment and rehabilitation of mined-out pits, and that there were ongoing discussions regarding an energy proposal to which the GOJ still has not responded. The company cited that over the eight-year period ending 2018 it spent US $1.2 million on land reclamation and restoration activities and a further US $1.9 million in 2017 on asset preservation. Under the circumstances the company indicated that the Minister's action was illegal and unlawful and that he misdirected himself as to what constitutes work. The company stated that the decision is a breach of the legitimate expectation to 'enjoy the benefits of the Kirkvine lease for a further 13 years' (Wilson and Barrett 2018). The matter was heard between March 19, 2019, and June 2020, and up to mid-2023 the Supreme Court still did not make its ruling.

With the invasion of Ukraine on February 24, 2022, the USA, UK, and EU had adopted sanctions aimed at freezing assets of the Russian president, select individuals, the Russian central bank, and many other companies. In Jamaica numerous groups such as the staff of WINDALCO , and their union were on edge about the possible US sanctioning of

Deripaska affecting UC RUSAL and the impact on WINDALCO. In March the UK had sanctioned Deripaska. The company however, reassured its employees that it does not expect to be sanctioned and further reminded them that UC RUSAL is managed by an independent board of directors, comprising citizens of various countries including UK and USA, and that the Board Chairman is a citizen of the Netherlands (Jackson 2022; Laidley 2022; Truss and Johnson 2022).

Since the invasion most of the larger shipments of alumina from Ewarton were sent to Russia. Previously, Anglo-Australian mining giant Rio Tinto Group and Anglo-Swiss metals miner Glencore Plc were major purchasers of RUSAL's Jamaican alumina for their North American facilities. With the invasion both companies applied self-sanctions and had ceased purchasing alumina from Ewarton. In March 2022 Glencore immediately began a review of their business with the intention to sell the 10.5% shares in En+ Group. Rio Tinto also announced at that time that it would sever ties with Russian businesses. In 2021, about 440,000 tonnes of alumina were sold, with the bulk going to Glencore's facility in the USA, and Rio Tinto's smelter in Quebec. Between March and April 2022, approximately 101,000 tonnes of alumina were shipped from Jamaica to Russia as other global suppliers such as in Ukraine and Australia had stopped (Al Insider 2022a; Al Insider 2022b Mines and Geology Division Production Report).

Towards the latter half of 2022 the Ewarton refinery faced shutdown resulting from a spillage of highly alkaline liquid draining from the effluent holding pond. The pond stores alkaline solutions ('technical water') from the Bayer process for recycling back into the plant. In recent years there were incidents of spillage from the pond which polluted the Rio Cobre River and induced environmental, social, and economic losses. At the beginning of 2022 the company had broken ground for a US $4 million (J$620 million) project to construct a second holding pond to prevent future spillage (Laidley 2022; Wilks 2022). However, continuous torrential rain in the summer resulted in the control system for drainage from the main pond being overwhelmed and caustic solutions overflowing and entering the Rio Cobre (river). This spillage resulted in large-scale death to aquatic life and the shutdown of many other activities downstream that relied on the river's water supply. The local environmental regulator notified the company that the operating permits would be suspended and that mining activities and the refinery should stop. The company was required to install the relevant systems to prevent future spillage, and that their environmental performance bond of over US $771,000 (J$117.76 million) is to be deducted to help in the recovery (Mattison 2022). As a result, the company expedited the construction of the second pond to temporarily hold the fluids and thus preventing another overflow. This system of recycling technical water had been installed by Alcan, the original builders of the Ewarton plant.

Following the departure of Alcan from Jamaica 2001, Kaiser followed soon in 2004. The parent Kaiser Aluminum and Chemical Corporation faced serious financial problems arising from labour disputes, pay-outs for asbestos claims, and energy problems in the USA. This led to the company filing for bankruptcy in 2002 and utilizing Chapter 11 of the federal bankruptcy code for reorganization. It also strategically disposed of its Jamaican operations. In April 2004, its 49% stake in the Discovery Bay, St. Ann facilities held by the subsidiary Kaiser Jamaica Bauxite Company, was sold to Century Aluminum Company and Noranda Aluminium Holding Corporation. The Government of Jamaica maintained its 51% share. Both Century and Noranda had equal share and jointly operated the mines in Jamaica and the Gramercy refinery in Louisiana. Century was formed by Glencore in 1995 to manage its aluminium producing assets. Noranda at that time was a subsidiary of Xstrata AG, an Anglo-Swiss multinational mining company based in Switzerland. The new company was called St. Ann Jamaica Bauxite Limited (Porter 2012a).

Kaiser Aluminum and Chemical Corporation, before filing bankruptcy, was the largest industrial employer in Spokane, Washington. It employed 2,000 workers but after seeking protection under the Chapter 11 bankruptcy code the number fell to less than 600. In the USA the company was impacted by prolonged labour disputes where employees were locked out of the company, which was deemed illegal in the courts. The West Coast energy crisis in the early 2000s had resulted in shut down of the Tacoma smelter in mid-2001. Three hundred workers who on average received annually US $35,000 salary became idle with no pay. While reorganizing, the company had to make payments of US $100 million in liability claims for asbestos manufactured and utilized by Kaiser. Kaiser Aluminium, having annual revenues of US $1.1 billion, was downsized to a business involved in fabrication for industrial markets. Towards the end of the 2000s, it focused on supplying Airbus and Boeing with high-grade aluminium for aerospace applications. It had disposed of its aluminium soft drink can businesses which were turning low profits (Gerianios 2002).

In May 2007 Noranda Aluminium Holding Corporation was acquired by New York-based private equity firm Apollo Global Management LLC, which had taken over some of Xstrata's aluminium assets. By August 2009 Century's share of St. Ann Jamaica Bauxite was sold to Noranda and so it became the sole owners of both the bauxite mines at St. Ann and the Gramercy refinery. In Jamaica the company was renamed Noranda Jamaica Bauxite Partnership where Noranda had full control of the 49% stake. The company having bauxite mining capacity of 4.8 million t/yr exported ore to Gramercy which has a 1.2 million t/yr alumina production capacity, and to the Sherwin alumina refinery at Corpus Christi, Texas (Porter 2012a). The Sherwin refinery was operational since 1953 but closed in mid-2016 by its owner Glencore after experiencing over 18 months of labour dispute and having accumulated a US $95 million debt (Freeman 2016).

Noranda Aluminium Holdings Corporation declared bankruptcy in 2016. This led to a US court of arbitration, in October 2016 approving New Day Aluminum LLC, affiliate of the Florida-based investment and management company DADA Holdings LLC as the preferred bidder for the purchase of Noranda's assets. These assets included the 49% stake in the Noranda Jamaica Bauxite Partnership in Jamaica, and the Gramercy refinery, Louisiana, USA which New Day acquired for US $24.2 million (Anonymous 2021). Following the sale and the transfer of Noranda Jamaica Bauxite Limited's assets to New Day, the Jamaican Government initially signed a binding Letter of Intent (LOI) with New Day in October 2016 for the company to officially begin mining. This LOI contemplated that a detailed agreement would be concluded by November 30, 2016. This was delayed for several reasons including the transitioning of the staff, a reopening of negotiations with the Government, and discussions on the boundaries of the Cockpit Country which heightened New Day's concerns about accessing the bauxite reserves. Further, frost in Florida had caused the pipes at Gramercy to burst causing financial losses (Montague 2018; Francis 2022).

The Government had opted for a profit-sharing scheme instead of the traditional bauxite levy arrangement. In this profit-sharing scheme both partners benefitted from any 'financial recovery' in the performance of the mining and refining operations. The Government stood to gain 17.33% of the earnings before interest, taxes, depreciation, and amortization (EBIDTA), or US $1.50 per tonne of dried bauxite shipped, whichever is greater. The standard royalty payments of US $0.50 per dry tonne of ore shipped remained in effect. This new arrangement was expected to earn the Government more in US dollars compared to the traditional production levy system. However, in early 2022, the Ministry of Finance indicated that the arrangement was underperforming and had netted less than the levy. The finance minister reported that from 2016 when the arrangement started until the end of 2021 when it was discontinued, Jamaica earned US $20 million from the profit share, of which US $9 million was still owing. He admitted that if the bauxite production levy had remained in effect the earnings would have been significantly greater at US $81.53 million or over J$12 billion (Barrett 2018; Montague 2018; Francis 2022).

The boundary for the Cockpit Country Protected Area was a matter of contention between various parties who could not agree on what specifically should be used to define the region. Various studies were conducted considering geomorphology (cockpit karst) and ecology and all had contrasting views of the demarcation. The discussions took about ten years until November 21, 2017, when the Government finally announced the preliminary boundary that would be adopted to define the area. The Prime Minister Andrew Holness made the announcement and indicated that the designation was subject to ground truthing. He also declared that no mining would be permitted in the designated area, and this resulted in the loss of access to between 48 to 62 million tonnes of bauxite having value of approximately US $1.44 billion to US $1.85 billion, primarily within Trelawny. He further indicated that consistent with the declaration and the Mining Act, any existing mining leases would be amended. Consequently, adjustments were made to the bauxite exploration licence held by New Day (Holness 2017; Williams-Raynor 2017; Silvera 2020).

On September 6, 2018, New Day's agreement with the Government was finally signed, and the partnership was called Noranda Jamaica Bauxite Partnership II. At the ceremony New Day paid US $5 million for profit share and US $2 million against a US $12.6 million debt on the bauxite production levy inherited from Noranda. The Government was expected to make available 130 million tonnes of bauxite, and so the mining minister symbolically signed their new Special Mining Lease no. 173 (SML 173) as being 'approved in principle'. This ore field now provided access to 80 million tonnes of bauxite, and had originally spanned into the Cockpit Country but was modified to conform to the new designation. New Day, however, was not immediately able to utilize the SML 173 until an environmental permit for mining was obtained. However, until the permit was received, the company was able to mine bauxite from two other approved ore fields nearby (Barrett 2018; Montague 2018).

New Day faced prolonged delays in getting the environmental permit despite providing the relevant impact assessment from mid-2019. The impact assessment underwent numerous reviews until it was finally accepted and the environmental regulator, at an impromptu meeting on January 3, 2022, announced the release of a small area of the ore field which gave the company access to approximately 12 million tonnes of bauxite, the mining of which would be subject to numerous conditions. Of note, in March 2020 the mining lease was further downsized or 'clawed back' in the northwest by the Government. This buffer from the Cockpit Country denied access to 20 million tonnes of bauxite (Knight 2022). As an alternative, the Government provided the company with resources in St. Ann to make up for that denied. This alternative ore field was one of the six leased by Alcoa in the late 1960s but was never worked (Hill and Williams 1974). On March 18, 2022, the Cockpit Country Protected Area of 78,024 hectares (ha) was made law (Government of Jamaica 2022a). By notice under the Mining Act, the area was declared closed to prospecting and mining (Government of Jamaica 2022b). The area clawed back by the Government in 2020 was also included within the boundary, despite not being encompassed in the 2017 declaration, and not being cockpit karst or of ecological sensitivity. This represented an additional US $600 million of bauxite sterilized.

In 2021 New Day sold its Jamaican bauxite and American alumina assets to London-based Concord Resources Ltd. Concord the new managing partner, renamed their Jamaican company Discover Bauxite Operations Ltd, and the partnership became Discovery Bauxite Partners. Concord had accepted the liabilities of New Day and inherited the ongoing matters pertaining to the access to bauxite reserves. The GOJ began negotiations with Concord on the fiscal regime which, starting in 2022 would include retiring the profit-share arrangement and adopting the standard production levy. Negotiations, however, were suspended due to the limited permitting of the mining lease (Knight 2022) and the Government seeking to provide the company with alternate resources. The Government and Concord were also sued by various groups on claims of breaching their Constitutional rights. Further, the claimants independently sought injunctions from the Supreme Court to block mining in the mining leases held by the company. The matters were heard throughout 2022 and are scheduled to be continued in November and December 2023 (Anonymous 2022; Gayle and Johnson 2022; Mundle 2022; Supreme Court of Jamaica 2022).

When Kaiser sold the shares in the Discovery Bay mines in 2004 it also disposed of its holdings in Alumina Partners of Jamaica (Alpart) in St. Elizabeth. This operation boasted 1.65 million t/yr or 38% of the alumina capacity in Jamaica. Before the sale, the partnership comprised Kaiser having 65% and Oslo-based Norsk Hydro ASA 35%. Norsk Hydro was given first preference to Kaiser's stake and followed through in acquiring it. However, the company shortly sold it to Glencore. At the time, UC RUSAL, the second largest aluminium producer made attempts for the acquisition but failed. When UC RUSAL incorporated Glencore's alumina facilities in 2007, it also obtained the majority shares in Alpart. By September 2011, however, Norsk Hydro sold its 35% stake to UC RUSAL which became the sole owner of the company and majority owner of all the bauxite and alumina interests in Jamaica. The acquisitions of both WINDALCO and Alpart by UC RUSAL represented 65% of the total alumina capacity in the island (Porter 2012a). UC RUSAL had shut down Alpart in 2009, but temporarily restarted mining in 2015 to temporarily export bauxite to a refinery in Russia. In June 2016, UC RUSAL sold Alpart to JISCO who reopened the refinery in October 2017 and exported its first shipment of 35,000 tonnes of alumina to China on December 29, 2017 (Chinese Embassy 2018).

The Gansu Province of China which owns JISCO proposed a US $6 billion investment to build and operate a special economic zone and industrial park over 27 km^2 in St. Elizabeth, centred on the alumina plant. This mega-project was called the Jamaica JISCO International Industrial Park Development Plan and involved the development of new industries and production lines. This included upgrading the existing refinery from 1.65 to 2 million t/yr, constructing a new 2 million t/yr alumina refinery, building a new 80,000 t/yr aluminium sheet mill, and establishing a special economic zone having special economic regulations that differ from general trade, tax, and investment rules, to attract other businesses (Anonymous 2018; Myers 2018).

Work began with the upgrading of the Alpart plant. This however was not without challenges as the new owners had to overhaul the aged plant infrastructure which was nearing 50 years old. The plant was shut down for eight years, and the equipment and piping were corroded. The plant had general low efficiency and high energy consumption, and this resulted in frequent malfunctions. The cost of production had reached over US $480 per tonne at a time when alumina prices were trending less than US $300 per tonne. After discussions with the Government, JISCO announced that the operations at the plant would be suspended for two years to facilitate the US $1.1 billion modernization and expansion. The owners also had difficulties sourcing parts for the old equipment as some manufacturers had discontinued production of certain parts and stopped providing technical support to the systems used. Since acquiring the plant the company had expended US $300 million on the initial rehabilitation and upgrades. The closure of the plant resulted in most of the over 950 workers being laid off (Fastmarkets 2019; Henry 2019; Linton 2019). The upgrades were further stalled by the COVID-19 pandemic which limited the ability of the Chinese to focus on the operations. Further, the delays by the company to acquire the lands having the necessary bauxite reserves to support the upgraded refinery capacity have led to widespread sterilization of the orebodies. The Chinese Government also required that the company demonstrate that reserves are available before approval is given for funding to support the upgrade and expansion works (JISCO Alpart Jamaica 2022). At the close of 2022, the restart of Alpart remained uncertain.

Apart from the UC RUSAL-owned alumina operations, bauxite refining was also done at JAMALCO, a joint venture between 55% shareholder Alcoa Minerals of Jamaica and the Government-owned Clarendon Alumina Production (CAP) which has 45% share. Alcoa Minerals of Jamaica is a subsidiary of Alcoa World Alumina and Chemicals, the parent which operates Alcoa's refineries. In 2011, Jamaica had 4.335 million t/yr nameplate alumina production capacity which included the 1.26 million tonnes from WINDALCO, 1.65 million tonnes from Alpart and 1.425 million tonnes capacity of JAMALCO. UC RUSAL maintained 93% interest in WINDALCO and full interest in Alpart. JAMALCO however had both Jamaican and North American ownership (Porter 2012b).

The JAMALCO joint venture was formed in 1976 when the Government increased state participation in the companies. The Government, through JBM, owned 6% and Alcoa Minerals Jamaica Inc had 94%. Alcoa closed the plant in 1985, but the Government reopened it and successfully managed to operate it to increased productivity. Alcoa returned in 1988, and under an agreement with CAP, Alcoa sold CAP a 44% interest. The Government therefore at that time had 50% in total, represented by both CAP and JBM. However, CAP later acquired JBM's share. In 2007, however, CAP's holdings were reduced to 45% when Alcoa solely funded an expansion of the alumina plant capacity which entitled it to a greater share of the alumina produced.

Due to declining revenue from the export of bauxite and alumina, the Government had entered a US $1.3 billion loan agreement with the IMF to support the budget early in 2010. The IMF had advised that to meet its fiscal target, the Government should sell its 45% equity stake in the JAMALCO joint venture. The Government contemplated this, and by April it announced an agreement with Chinese company Zhuhai Hongfan to complete the acquisition. This, however, was pending Alcoa's decision to take up the offer (Porter 2012a). By early 2011 it was revealed that negotiations with the Chinese had ceased, and discussions were being held with Glencore International AG. Such a deal would be conflicting as Glencore would be both Alcoa's partner and competitor. Glencore at that time had an 8.8% share in UC RUSAL, the majority shareholder in the Jamaican bauxite industry. It was disclosed that both companies had agreed for UC RUSAL to take over Glencore's portion, and that UC RUSAL had insisted on having preference to the CAP shares in return for opening the Kirkvine works. Such a situation would have resulted in UC RUSAL controlling 82% of the island's alumina production capacity (Porter 2012a).

In mid-2013, CAP entered a prepayment arrangement with Singapore-based Noble Resources Ltd, a member of the Hong Kong-based Noble Group Holdings Ltd (Noble), where Noble Resources provided US $120 million at 8% interest to CAP to repay working capital debts to Alcoa for the JAMALCO joint venture and to fund any shortfalls that were expected during the depression of the world alumina market. There was also an alumina sales agreement where CAP provided alumina to Noble Resources, but also giving them the option to acquire CAP and it stake in JAMALCO. The arrangement with Noble Resources was made due to Jamaica's 'bailout' agreement with the IMF that required Jamaica to either sell CAP or enter into an asset lease agreement by December 2013. The Government had previously negotiated with Glencore, but that company's proposal was not deemed acceptable. CAP had a long-standing forward sales agreement with Glencore from 2002 which committed the provision of 600,000 tonnes of alumina under fixed-price terms that resulted in Jamaica losing approximately US $400 million. The fixed-price terms of that agreement ended in 2012 and freed up a third of CAP's alumina allotment which represented the 200,000 tonnes being provided to Noble Resources. Even though CAP's agreement with Glencore ended in 2015, the latter sought to block the deal with Noble Resources. Glencore sought an ex-parte injunction, but the Government won the court challenge. As such, CAP was free to supply Noble Resources and the first shipment of alumina departed Rocky Point, Clarendon, at the end of July 2013 (Anonymous 2013).

Alcoa's tenure in Jamaica ended in December 2014 when its 55% shares in JAMALCO were sold to Noble for US $132.7 million. Noble, however, had estimated the fair value of the stake at US $289 million. The sale of JAMALCO for Alcoa represented a US $100 million non-cash loss, but the company expressed that the sale was in line with their global strategy to reshape its upstream portfolio and lower the cost base for their commodity business. The sale included an agreement that Alcoa would continue as the managing operator for three years under a compensated service agreement and the employees would be retained. At the time of sale, the refinery's capacity was 1.425 million t/yr of alumina and the sale provided Noble with access to an additional 778,800 t/yr alumina. The Government, through CAP, retained its 45% interest and access to 646,000 t/yr of alumina (Anonymous 2014; Farchy 2017). The Government had also scrapped the plan to divest its share in the refinery, as was originally suggested by the IMF to help meet fiscal targets. It was agreed that the Government would put no more money into the joint venture, as Noble was willing to upfront the capital requirements (Luton 2014b).

In December 2017, Noble assumed the role of managing partner in the JAMALCO joint venture. The Government's debt from the prepayment facility continued to grow and in mid-2020 this amounted to US $135.7 million which comprised the principal of US $120 million and accrued interest. The refinery assets at that time were estimated to value US $477 million. The Government paid off the debt in full in September 2020 in anticipation of listing JAMALCO on the Jamaica Stock Exchange. Jamaica also sought to exit the alumina supply agreement which expires in 2025. CAP would instead enter a three-year marketing agreement that would require Noble to sell CAP's alumina share for the same price they receive on the international market and receive a commission (Anonymous 2020a, b).

JAMALCO managed to double its revenue from US $6.8 million in the first half of 2020, to US $16 million in the first half of 2021. Unfortunately, on August 22, 2021, failure of a flange on a fuel line at the company's powerhouse resulted in a devastating fire. This stopped production for over ten months and severely impacted the company's revenue. For the period of January to September 2021 Noble reported a net loss of US $67 million, and for the same period in 2022 the loss was US $97 million. The company, however, was insured up to US $250 million and relied on pay-out from the policy to rebuild the operations. In 2022 the insurance payments amounted to US $86 million. With the significant losses incurred from the fire, Noble, in early 2022, began contemplating the disposal of its stakes in JAMALCO. In September 2022 Noble reported net debt of US $495 million (Anonymous 2023). Back in 2017, in the midst of a crisis, Noble had also considered selling the refinery to restructure its US $3.5 billion debt (Iqbal 2017).

By the summer of 2022, JAMALCO restarted production to half capacity. In October 2022 the company reported that it was back at full production and had exceeded production targets. The full reconstruction is projected to be completed by 2024. However, despite returning to full production, the company struggled with the cost of steam. Back in 2020 the company, under an agreement, started receiving steam from an adjacent newly commissioned, natural gas-fired combined heat and power plant. This plant is privately owned and was constructed at a cost of US $285 million. The fire did not affect this power plant but resulted in an interruption of the steam supply agreement which led to a new agreement being made where steam is sold at a higher price. JAMALCO indicated that the high price of steam was due to the ten-fold increase in the price of natural gas over the previous year which reflected the effect of price hikes in Europe due to the Russia-Ukraine war. At the end of 2022 JAMALCO was considering energy solutions using heavy fuel oil, and the company had already arranged for shipments of this commodity (Jackson 2021a, b; Collinder 2022; Jackson 2022; JAMALCO 2022). By March 2023 it was announced that Noble's 55% share in JAMALCO would be sold. On May 2, 2023, US-based producer Century Aluminum Company acquired the shares and declared that it had paid US $1 for them. The details of the sale were undisclosed, but this 'low price' seems to suggest that Century assumed liabilities of Noble which was "experiencing financial distress following curtailment of JAMALCO's operations . . . in 2021 due to a facility fire". Century also announced plans to bring the plant to full production capacity in 2024 by investing up to US $20 million, as in mid-2023 it was operating at 80% capacity, or 1.2 million t/yr. Glencore has 42.9% ownership of Century (Century Aluminum 2023; Jackson 2023a, b).

References

AAA (1978) Revere Copper and Brass, Inc. v, Overseas Private Investment Corporation. American Arbitration Association (AAA), Case no. 1610013776, 24 Aug 1978

Al Circle (2014) Paulwell announces Rusal can restart mining at Alpart Jamaica. Al Circle Newsletter, 24 December 2014. https://www.alcircle.com/news/paulwell-announces-rusal-can-restart-mining-at-alpart-jamaica-20465. Accessed 8 Dec 2022

Al Insider (2022a) Glencore mulling sale of its share in Rusal's parent company En+ Group. Aluminium Insider, 4 March 2022. https://aluminiuminsider.com/glencore-mulling-sale-of-its-share-in-rusals-parent-company-en-group/. Accessed 6 Mar 2022

Al Insider (2022b) Rio Tinto to sever ties to Russian businesses due to Ukraine invasion. Aluminium Insider, 12 March 2022. https://aluminiuminsider.com/rio-tinto-to-sever-ties-to-russian-businesses-due-to-ukraine-invasion/. Accessed 14 Mar 2022

Angus GL (2020) Mining Communities to benefit from $60 Million under BCDP. Jamaica Information Service, 14 Oct 2020. https://jis.gov.jm/mining-communities-to-benefit-from-60-million-under-bcdp/. Accessed 4 Oct 2023

Anonymous (1946) Mining Policy for Colonies. Daily Gleaner, 10 Dec 1946, p 1 & 16

Anonymous (1974) Final act of the international conference of bauxite producing countries. International Bauxite Association, Conakry

Anonymous (1975) At bauxite-alumina plant in Maggotty, Revere Suspends Operations. The Daily Gleaner, 20 Aug 1975, p 1

Anonymous (1998) Revere Ware Corporation History. International Directory of Company Histories 22:np. http://www.fundinguniverse.com/company-histories/revere-ware-corporation-history/. Accessed 30 Nov 2011

Anonymous (2011) Rusal, Jamaica stand-off. The Daily Gleaner, 25 Mar 2011. https://jamaica-gleaner.com/article/business/20110325/rusal-jamaica-stand. Accessed 8 Dec 2022

Anonymous (2013) Noble has option to buy. The Daily Gleaner, 7 Aug 2013. https://jamaica-gleaner.com/gleaner/20130807/business/business1.html. Accessed 7 Dec 2022

Anonymous (2014) Alcoa selling JAMALCO stake to Noble Group. Engineering and Mining Journal, Dec 2014. https://www.e-mj.com/news/latin-america/alcoa-selling-JAMALCO-stake-to-noble-group/. Accessed 7 Dec 2022

Anonymous (2018) $6billion industrial park and special economic zone earmarked for Nain, St Elizabeth. Jamaica Gleaner, 11 Feb 2018. https://jamaica-gleaner.com/article/news/20180211/6billion-industrial-park-and-special-economic-zone-earmarked-nain-st-elizabeth. Accessed 21 Mar 2023

Anonymous (2020a) CAP to pay off debt in lay up to JAMALCO IPO. The Daily Gleaner, 14 Aug 2020. https://jamaica-gleaner.com/article/business/20200814/cap-pay-debt-lay-JAMALCO-ipo. Accessed 7 Dec 2022

Anonymous (2020b) Jamaica clears debt owed to alumina partner Noble Group. The Daily Gleaner, 9 Sept 2020. https://jamaica-gleaner.com/article/business/20200909/jamaica-clears-debt-owed-alumina-partner-noble-group. Accessed 7 Dec 2022

Anonymous (2021) Sold! Jamaica observer. 16 Jul 2021. https://www.jamaicaobserver.com/business/sold/. Accessed 21 Mar 2023

Anonymous (2022) Bauxite battle, discovery bauxite embroiled in court as opponents seek to shutter operations. Daily Observer, 7 Dec 2022. https://www.jamaicaobserver.com/business/bauxite-battle/. Accessed 10 Dec 2022

Anonymous (2023) Jamalco set for sale. Jamaica Gleaner, 17 Mar 2023. https://jamaica-gleaner.com/article/business/20230317/jamalco-set-sale. Accessed 21 Mar 2023

Barclay LA, Girvan N (2013) Transnational restructuring and the Jamaican bauxite industry: the swinging pendulum of bargaining power. In: Gendron RS et al (eds) Aluminum ore: the political economy of the global bauxite industry. UBC Press, Vancouver, pp 238–267

Barker FG (1971) Remarks on the evolution of the alumina segment of the aluminium industry. J Geol Soc Jam Bauxite/alumina Symp 1:52–57

Barrett L (2018) Bad deal! Opposition knocks new profit-sharing bauxite agreement between Jamaica and US company. Jamaica Gleaner, 18 Sept 2018. https://jamaica-gleaner.com/article/lead-stories/20180919/bad-deal-opposition-knocks-new-profit-sharing-bauxite-agreement. Accessed 21 Mar 2023

Beard EH (ed) (1954) Reports from the Geological Surveys—Jamaica. Col Geol Min Res 4(1):81–83

Bell EO (1979) Capital Development Fund—Annual Report 1978. Ministry Paper No. 24. Ministry of Finance (Jamaica), Kingston. 7 June 1979. https://nlj.gov.jm/MinistryPapers/1979/no.%2024.pdf. Accessed 18 Mar 2023

Brown K, Reckord M (1970) The Alpart story. Alpart Public Relations Department, Jamaica

Buchholz E (1952) Alumni notes—Schmedeman, Otto. C. Wisconsin Engineer 57(2):21

CBC (2007) Rio Tinto reaches $38.1B US merger deal with Alcan. CBC News, 29 Mar 2012

Century Aluminum (2023) Century Aluminum Company, Form 10-Q, Quarterly Report Pursuant to Section 13 or 15(d) of the Securities Exchange Act of 1934, For the quarterly period ended June 30, 2023. United States Securities and Exchange Commission, Washington DC. https://centuryaluminum.com/investors/filings-and-financials/quarterly-results/default.aspx. Accessed 5 Oct 2023

Charles II (1661) A proclamation for the encouraging of planters in His Majesties island of Jamaica in the West Indies (Whitehall, December 14, 1661). John Bill and Christopher Barker (Printers to the King's Most Excellent Majesty), London

Chin LAD (1971) The effects of important properties of Jamaican bauxites on the Bayer process. J Geol Soc Jamaica, Bauxite/alumina Symp I:58–61

Chinese Embassy (2018) First shipment of alumina from JISCO Alpart Jamaica. Embassy of the People's Republic of China in Jamaica. 3 Jan 2018. http://jm.china-embassy.gov.cn/eng/news/202209/t20220902_10761410.htm. Accessed 10 Dec 2022

Chubb LJ (1961) Establishment of the Mines Department. Geonotes IV(3 & 4):18

Clarke H (1979) Agreement with bauxite/alumina industry. Ministry Paper No. 46, File no. GCR/9, Ministry of Mining and Natural Resources (Jamaica), Kingston, 25 Oct 1979. http://www.nlj.gov.jm/MinistryPapers/1979/no.%2046.pdf. Accessed 20 Mar 2023

Collinder A (2014) Government to wind up BATCo. Jamaica Gleaner, 23 Jul 2014. https://jamaica-gleaner.com/gleaner/20140723/business/business2.html. Accessed 22 Mar 2023

Collinder A (2022) JAMALCO back to production. The Daily Observer, 7 Sept 2022. https://www.jamaicaobserver.com/business/JAMALCO-back-to-capacity/. Accessed 7 Dec 2022

Cover WA (ed) (1953) The Handbook of Jamaica for the year 1953. Government Printing Office, Kingston

Cromwell O (1655) By the Protector, a Proclamation giving encouragement to such as shall transplant themselves to Jamaica (Whitehall, October 10, 1655). Henry Hills and John Fields (Printers to His Highness), London

Cross B (2013) White metal: bauxite, labour, and the land under Alcan in twentieth-century Guyana, Jamaica, and Australia. In: Gendron RS et al (eds) Aluminum ore: the political economy of the global bauxite industry. UBC Press, Vancouver, pp 302–327

Das S, Padel F (2013) Battles over bauxite in East India: the Khondalite mountains of Khondistan. In: Gendron RS et al (eds) Aluminum ore: the political economy of the global bauxite industry. UBC Press, Vancouver, pp 328–352

Davis CE (1974) Alternatives to bauxite. Jam Jour 8(2 and 3):53–55

Davis CE (1989) Jamaica in the world aluminium industry 1938–1973, Volume I. Jamaica Bauxite Institute, Kingston

Davis CE (1995a) The Jamaican bauxite industry: glimpses into its past, present, and future. In: Miller RL, Escalante G, Reinemund JA, Bergin MJ (eds) Energy and Mineral Potential of the Central American-Caribbean region. Circum-pacific council for energy and mineral resources earth science series, Springer, Berlin, Heidelberg, 16, pp 315–318. https://doi.org/10.1007/978-3-642-79476-6_42

Davis CE (1995b) Jamaica in the world aluminium industry, Volume II 1974—1988, Bauxite levy negotiations. Jamaica Bauxite Institute, Kingston

Davis CE (2000) Reynolds Metals Company 1928–2000. Daily Gleaner, 28 May 2000. http://jamaica-gleaner.com/gleaner/20000528/Business/Business5.html. Accessed 22 Nov 2011

Davis CE (2011) Jamaica in the world aluminium industry, Volume III, The partnership and joint venture agreements. Jamaica Bauxite Institute, Kingston

Davis G (2015) UC Rusal announces full resumption of operations at Alpart. Jamaica Information Service, 8 Sept 2015. https://jis.gov.jm/uc-rusal-announces-full-resumption-of-operations-at-alpart/. Accessed 8 Dec 2022

Farchy J (2017) What's left for Noble Group creditors: the answer's in Jamaica. Bloomberg, 19 Nov 2017. https://www.bloomberg.com/news/articles/2017-11-19/why-noble-group-s-debt-restructuring-may-have-a-jamaican-flavor#xj4y7vzkg; and The Straits Times (Reproduced from Bloomberg; Open access), 20 Nov 2017. https://www.straitstimes.com/business/companies-markets/whats-left-for-noble-group-creditors-the-answers-in-jamaica. Accessed 21 Mar 2023

Fastmarkets (2019) LME WEEK 2019: aluminium fundamentals and forecast. Fastmarkets, 25 Oct 2019. https://www.fastmarkets.com/insights/lme-week-2019-aluminium-fundamentals-and-forecast. Accessed 10 Dec 2022.

Francis K (2022) Robinson, Clarke spar over missed bauxite earnings. Jamaica Gleaner, 19 Jan 2022. https://jamaica-gleaner.com/article/lead-stories/20220119/robinson-clarke-spar-over-missed-bauxite-earnings. Accessed 21 Mar 2023

Freeman S (2016) Sherwin Alumina closes plant after 62 years. Corpus Christi Business News, 2 Aug 2016. https://www.ccbiznews.com/Sherwin-Alumina-closes-plant-after-62-years. Accessed 6 Dec 2022.

Gayle B, Johnson J (2022) Injunction sought to halt Noranda mining near Cockpit Country, hearing set for May 31. The Gleaner, 19 May 2022. https://jamaica-gleaner.com/article/news/20220519/injunction-sought-halt-noranda-mining-near-cockpit-country-hearing-set-may-31. Accessed 10 Dec 2022.

Gerianios NK (2002) Kaiser Aluminum files for Chapter 11 to reorganize. Seattle Pi, 12 Feb 2002. http://www.seattlepi.com/business/article/Kaiser-Aluminum-files-for-Chapter-11-to-reorganize-1080129.php. Accessed 10 Jul 2014

Gittleson K (2019) US lifts sanctions on Putin ally's firms. BBC News, 27 Jan 2019. https://www.bbc.com/news/business-47023004. Accessed 8 Dec 2022

Government of Jamaica (1953) Colonial reports, Jamaica 1951, or report on Jamaica for the year 1951. Her Majesty's Stationery Office, London

Government of Jamaica (1959) Annual Report of the Department of Mines for the Financial Year 1957–58. The Government Printer, Kingston

Government of Jamaica (1960) Annual Report of the Department of Mines for the Financial Year 1958–59. The Government Printer, Kingston

Government of Jamaica (1963) Annual Report of the Department of Mines for the Financial Year 1961–62. The Government Printer, Kingston

Government of Jamaica (2022a) The Natural Resources Conservation (Cockpit Country Protected Area) Order, 2022. The Jamaica Gazette Supplement, 18 Mar 2022, Jamaica Printing Services (1992) Limited, CXLV(21):343–349

Government of Jamaica (2022b) The Mining (Prohibition) (Cockpit Country Protected Area) Notice, 2022. The Jamaica Gazette Supplement, 18 Mar 2022, Jamaica Printing Services (1992) Limited, CXLV(19A):340A

Government Public Relations Office (1960) The Handbook of Jamaica for 1960. Government Printing Office, Kingston

Government Public Relations Office (1961) The Handbook of Jamaica for 1961. Government Printing Office, Kingston

Government of Jamaica (2004) The Bauxite (Production Levy) Act, Capital Development Fund (Withdrawal) (No. 1), Order, 2004. Jam Gaz, 18 March 2004, L.N.41, CXXVII(32):87–88

Government of Trinidad and Tobago (1949) Report to the Minister of Finance on the accounts of the territory of Trinidad and Tobago for the Year 1948 (Trinidad and Tobago Auditor General). Government Printer Office, London

Gow NN, Lozej GP (1993) Bauxite. Geosci Can 20(1):9–16

Henry B (2019) 2-year closure? Daily Observer, 11 Sept 2019. https://www.jamaicaobserver.com/news/2-year-closure/. Accessed 10 Dec 2022

Hill VG (1971) Preface (Bauxite/alumina symposium 1971). J Geol Soc Jam Bauxite/alumina Symp 1971:i

Hill VG, Williams SA (1974) The relationship between the Bauxite Industry and the Agricultural Sector in Jamaica. Mines and Geology Division, Ministry of Mining and Natural Resources, Kingston. https://ageconsearch.umn.edu/record/264437/files/wiae-1974-16.pdf. Accessed 18 Mar 2023

Holness A (2017) Statement in Parliament on the delimitation of the boundary of the Cockpit Country and the Cockpit Country Protected Area. The Office of the Prime Minister (Jamaica), Kingston, 21 Nov 2017, https://opm.gov.jm/speech/statement-in-parliament-on-the-delimitation-of-the-boundary-of-the-cockpit-country-and-the-cockpit-country-protected-area/. Accessed 10 Dec 2022

Hose HR (1950) The geology and mineral resources of Jamaica. Col Geol Min Res 1(1):11–36

Huggins HD (1965) Aluminium in Changing Communities. Andre Deutsch Limited, London

Iqbal H (2017) Noble Group mulls sale of Jamalco alumina refinery. AlCircle 21 Nov 2017. https://www.alcircle.com/news/noble-group-mulls-sale-of-jamalco-alumina-refinery-29212. Accessed 21 Mar 2023

Jackson S (2021a) JAMALCO fire recovery plan to cost $3b now, more later. The Daily Gleaner, 6 Oct 2021. https://jamaica-gleaner.com/article/business/20211006/JAMALCO-fire-recovery-plan-cost-3b-now-more-later. Accessed 7 Dec 2022

Jackson S (2021b) JAMALCO gets US $86m insurance payout. The Daily Gleaner, 9 Sept 2022. https://jamaica-gleaner.com/article/business/20220909/JAMALCO-gets-us86m-insurance-payout. Accessed 7 Dec 2022

Jackson S (2022) Rusal says Windalco unaffected by sanctions. The Daily Gleaner, 11 Mar 2022. https://jamaica-gleaner.com/article/business/20220311/rusal-says-windalco-unaffected-sanctions. Accessed 8 Dec 2022

Jackson S (2023a) Jamalco sold for US$1. The Gleaner, 26 April 2023. https://jamaica-gleaner.com/article/business/20230426/jamalco-sold-us1. Accessed 4 Oct 2023

Jackson S (2023b) Century books $16b gain on Jamalco deal. Financial Gleaner, 11 Aug 2023. https://jamaica-gleaner.com/article/business/20230811/century-books-16b-gain-jamalco-deal. Accessed 4 Oct 2023

Jamaica Information Service (1965) The Handbook of Jamaica for 1965. The Government Printing Office, Kingston

JAMALCO (2022) JAMALCO recorded excellent performance in October. JAMALCO NewsLink, 25 Nov 2022, 16(11):1–2

JBI (2021a) The JBI's alumina process pilot plant. Jamaica Bauxite Institute, Kingston. https://jbi.org.jm/pilot-plant/. Accessed 24 Mar 2023

JBI (2021b) Bauxite Community Development Programme. Jamaica Bauxite Institute, Kingston. https://jbi.org.jm/bcdp/. Accessed 24 Mar 2023

Jiang M (2011) Rusal to increase stake in Jamaica's Alpart to 100%. Shanghai Metals Market, 18 Sept 2011. https://news.metal.com/newscontent/100023988/Rusal-To-Increase-Its-Stake-In-Jamaicas-Alpart-To-100/. Accessed 8 Dec 2022

JIS (2007) Russian owners have expansion plans for Alpart, Windalco. Jamaica Information Service, 14 Jun 2007. https://jis.gov.jm/russian-owners-have-expansion-plans-for-alpart-windalco/. Accessed 8 Dec 2022

JISCO Alpart Jamaica (2022) Environmental impact assessment, proposed mining and quarrying at "Outer Valley" Section of SEPL 541, St. Elizabeth. February 2022, Enviroplanners Ltd. https://www.nepa.gov.jm/sites/default/files/2022-02/EIA%20Final%20Draft%20Report%20Outer%20Valley%20Section%20SEPL%20541%20-%20Feb%202022.pdf. Accessed 21 Mar 2023

JNARDDC (2018) About JNARDDC. Jawaharlal Nehru Aluminium Research Development and Design Centre, India. https://jnarddc.gov.in/en/about_us/jnarddc.aspx. Accessed 24 Mar 2023

Jones GA (1952) The truth about bauxite discovery and development in Jamaica. Daily Gleaner, 20 Dec 1952, p 8

Knight P (2022) Stakeholder's briefing on NRCA decision of SML 173. National Environment and Planning Agency. https://www.nepa.gov.jm/sites/default/files/2022-01/Stakeholders%20briefing_FINAL_2%20January%202021_0.pdf. Accessed 10 Dec 2022

Laidley A (2022) Windalco workers worried. Jamaica Observer 2 Mar 2022. https://www.jamaicaobserver.com/business/windalco-workers-worried/. Accessed 8 Dec 2022

Lightbourne RC (1970) Mining. Ministry Paper No. 33/70, File no. 14/1/11. Ministry of Trade and Industry (Jamaica), Kingston, 8 Jun 1970. http://www.nlj.gov.jm/MinistryPapers/1970/33.pdf. Accessed 20 Mar 2023

Linton L (2019) JISCO/Alpart operations set to be suspended for up to two years. Jamaica Information Service, 11 Sept 2019. https://jis.gov.jm/jisco-alpart-operations-set-to-be-suspended-for-up-to-two-years/. Accessed 10 Dec 2022

Linton L (2023) House Approves Withdrawal from Capital Development Fund. Jamaica Information Service. 20 Sept 2023. https://jis.gov.jm/house-approves-withdrawal-from-capital-development-fund/. Accessed 4 Oct 2023

Luton D (2014a) Rusal's Ewarton mining licence also on the line—Paulwell, Rusal to meet today. The Daily Gleaner, 11 Jul 2014. https://jamaica-gleaner.com/gleaner/20140711/business/business1.html. Accessed 8 Dec 2022

Luton D (2014b) New direction at Jamalco—Jamaica shelves plan to sell CAP. The Daily Gleaner, 17 Oct 2014. https://jamaica-gleaner.com/article/business/20141017/new-direction-jamalco-jamaica-shelves-plan-sell-cap. Accessed 8 Dec 2022

Manley NW (1962) Alcoa minerals of Jamaica Inc. Ministry Paper No. 8 (MS13/48/01), Ministry of Development (Jamaica), Kingston, 27 Feb 1962. http://www.nlj.gov.jm/MinistryPapers/1962/No.8.pdf. Accessed 20 Mar 2023

Mattison R (2022) NEPA threatens to yank Windalco licence over Rio Cobre disaster. Jamaica Gleaner 6 Aug 2022. https://jamaica-gleaner.com/article/lead-stories/20220806/nepa-threatens-yank-windalco-licence-over-rio-cobre-disaster. Accessed 8 Dec 2022

Mines and Geology Division (1975) Annual Report for the Year ending 31st March 1975. Ministry of Mining and Natural Resources, Kingston

Montague R (2018) Statement on Novation Agreement by Robert Montague, Minister of Transport and Mining. Ministry of Transport and Mining Jamaica), Kingston, 19 Sept 2018. (Formerly accessible at www.mtw.gov.jm). Now accessible at https://mckoysnews.com/minister-montagues-statement-on-novation-agreement/. Accessed 21 Mar 2023

Mundle T (2022) Widow claims mining activities caused husband's death. The Gleaner, 1 Nov 2022. https://jamaica-gleaner.com/article/lead-stories/20221101/widow-claims-mining-activities-caused-husbands-death. Accessed 10 Dec 2022

Myers G (2018) Alpart expands. Jamaica Observer 1 Jan 2018. https://www.jamaicaobserver.com/central/alpart-expands/. Accessed 21 Mar 2023

Neita L (2009) Anecdotes from the bauxite sector. The Gleaner, 1 May 2009. http://www.jamaica-gleaner.com/gleaner/20090501/cleisure/cleisure3.html. Accessed 30 Nov 2011

Neita L (2022) Tell the whole story. Jamaica Observer, 9 Jan 2022. https://www.jamaicaobserver.com/columns/tell-the-whole-story-of-bauxite/. Accessed 30 Oct 2022

Porter ARD (2012a) What's next for bauxite? (Part I). The Gleaner, 19 Feb 2012. http://jamaica-gleaner.com/gleaner/20120219/focus/focus7.html. Accessed 10 Jul 2014

Porter ARD (2012b) What's next for bauxite? (Part II). The Gleaner, 20 Feb 2012. http://jamaica-gleaner.com/gleaner/20120220/cleisure/cleisure4.html. Accessed 10 Jul 2014

Reynolds-Baker A (2014) Government sells shares in Windalco to UC Rusal. Jamaica Information Service, 10 Jun 2014. https://jis.gov.jm/government-sells-shares-windalco-uc-rusal/. Accessed 8 Dec 2022

Rio Tinto (2007) Rio Tinto completes acquisition of 100% of Alcan. Rio Tinto Press Release, 15 Nov 2007

Roberts AL (1971) Mining of bauxite in Jamaica. J Geol Soc Jam Bauxite/alumina Symp 1:41–51

Richardson ER (1975) History of multinational corporations and bauxite producing countries. J Geol Soc Jam Proc Bauxite/alumina Symp 3:1–6

RUSAL (2007) European Commission approves RUSAL, SUAL and Glencore merger - creating world's number one aluminium company. RUSAL Press Release, 1 Feb 2007

RUSAL (2016) RUSAL completes the sale of Alpart alumina refinery in Jamaica. RUSAL Press Release, 22 Nov 2016

Sangster DB (1962) Report of the Jamaica Mission to Washington & London, June–July 1962. Ministry Paper No. 33, Ministry of Finance (Jamaica), Kingston, 23 Jul 1962 http://www.nlj.gov.jm/MinistryPapers/1962/No.33.pdf. Accessed 20 Mar 2023

Sawkins JG (1869) Reports on the geology of Jamaica, or Part II of the West Indian Survey, with contributions from G. P. Wall, Lucas Barrett, Arthur Lennox and C. B. Brown, and an appendix by R. Etheridge. (Memoirs of the Geological Survey), Longmans, Green, and Co, London

Silvera J (2020) Green light for bauxite mining in Cockpit Country. The Gleaner, 18 Nov 2020. https://jamaica-gleaner.com/article/lead-stories/20201118/green-light-bauxite-mining-cockpit-country. Accessed 10 Dec 2022

St. James Press (2000) International Directory of Company Histories—Alcan. http://www.fundinguniverse.com/company-histories/Alcan-Aluminium-Limited-Company-History.html. Accessed 22 Nov 2011

Supreme Court of Jamaica (2022) In the matter of the constitution of Jamaica, and in the matter of special mining leases permitting bauxite mining in areas where the claimants live and farm, and in the matter of an application for constitutional redress pursuant to Section 19 of the constitution. Supreme Court of Judicature of Jamaica in Civil Division, Claim No. SU 2022 CV 02353, [2022] JMSC Civ 218. https://supremecourt.gov.jm/sites/default/files/judgments/Grant%2C%20Victoria%20and%20Linsford%20Hamilton%20et%20al%20v%20Noranda%20Jamaica%20Bauxite%20Partners%2C%20New%20Day%20Aluminium%20%28Jamaica%29-%20Limited%20et%20al.pdf. Accessed 21 Mar 2023

Thompson M (2017) Appeal court upholds US $15m judgment against Western Cement. Daily Gleaner, 16 Aug 2017. https://jamaica-gleaner.com/article/business/20170818/appeal-court-upholds-us15m-judgment-against-western-cement. Accessed 6 Dec 2022.

Thomsen J (2019) Russian oligarch Deripaska loses U.S. court battle to lift sanctions. Reuters, 29 Mar 2022. https://www.reuters.com/legal/government/russian-oligarch-deripaska-loses-us-court-battle-lift-sanctions-2022-03-29/. Accessed 8 Dec 2022

Tretzel DC (1971) Development in world bauxite production and bauxite operations in Jamaica. J Geol Soc Jam Bauxite/alumina Symp I:28–37

Truss E, Johnson B (2022) Abramovich and Deripaska among 7 oligarchs targeted in estimated £15 billion sanction hit. GOV.UK Press Release. 10 Mar 2022. https://www.gov.uk/government/news/abramovich-and-deripaska-among-seven-oligarchs-targeted-in-estimated-15bn-sanction-hit. Accessed 8 Dec 2022

Vogel KP, Rappeport A (2019) Russian Oligarch sues the U.S. over sanctions. The New York Times, 15 Mar 2019. https://www.nytimes.com/2019/03/15/us/politics/oleg-deripaska-russia-sanctions.html. Accessed 8 Dec 2022

Wacaster S (2011) 2009 Minerals Yearbook, Islands of the Caribbean, The Mineral Industries of the Islands of the Caribbean. United States Geological Survey, U.S. Department of the Interior. https://d9-wret.s3.us-west-2.amazonaws.com/assets/palladium/production/mineral-pubs/country/2009/myb3-2009-aa-bf-bb-uc-jm-td.pdf. Accessed 18 Mar 2023

Walker HS (1971) Address on the bauxite/alumina industry of Jamaica. J Geol Soc Jam Bauxite/alumina Symp 1971:1–2

Wilks A (2022) Windalco banks on additional waste-holding pond to protect Rio Cobre. Jamaica Gleaner 7 Jan 2022. https://jamaica-gleaner.com/article/lead-stories/20220107/windalco-banks-additional-waste-holding-pond-protect-rio-cobre. Accessed 8 Dec 2022

Williams-Raynor P (2017) Gov't breaks silence on Cockpit Country. The Gleaner, 26 Apr 2017, p C9

Wilson HE (1985) Down to Earth: one hundred and fifty years of British Geological Survey. Scottish Academic Press Ltd., Edinburgh, p 1985

Wilson N, Barrett L (2018) Bauxite battle, Russians take Gov't to court for revoking licence. The Gleaner, 24 Jan 2019, p A1

Zans VA (1951) Economic geology and mineral resources of Jamaica. Bull Geol Surv Jam 1:32

Zans VA (1952) Bauxite resources of Jamaica and their development. Col Geol Min Res 3(4):307–333

5 The Modern Geological Survey

Abstract

Modern national Geological Surveys are tasked to collect, compile, and communicate information of the earth resources endowed to a nation. Further, the expertise of the Government geoscientist is required for the examination and disposal of all minerals, energy, and water resources of a country, for its economic benefit, for the health and safety of its inhabitants, and for the protection of the environment. The evolution of geological survey activities and geoscience research in Jamaica over the last 70 years is explored. The roles of various entities are highlighted to show their impacts on national development and on shaping Jamaica to what it is today.

5.1 A Survey for Economic Purposes

At a March 1943 sitting of the Legislative Council in Kingston, Rev John William Maxwell, representative of Trelawny parish, gave notice that at the next sitting he intended to move that a geological survey of the Jamaica should be conducted by the Government with the aim of discovering mineral resources (Anonymous 1943a). By September, outgoing Governor Sir Arthur Richards, in his report on the conditions of the island, indicated that the potential of the recent bauxite discoveries was not fully determined but were showing promise. He further mentioned that an application was made to Britain for assistance in conducting a thorough geological survey of the island. His report also mentioned the encouraging potential for copper and the intentions to start an island cement industry (which materialized nine years later in 1952) (Anonymous 1943b; Government of Jamaica 1956). It seems that the requirements for a thorough geological survey were not properly understood as suggested by the draft estimates of expenditures for the financial year 1944–1945 where a grossly inadequate £300 for a geological survey was allocated (Anonymous 1944).

5.2 Colonial Development and Welfare

Governor Richards in his review encouraged the colony to become financially independent of the British Treasury noting that the actual financial position of the island was masked by grants from the Colonial Development and Welfare (CD&W) Fund and other subsidies from the English Government (Anonymous 1943b). Between 1938 and 1939 a Royal Commission was established to examine the potential for development of the British West Indian Colonies (Anonymous 1947). The Second World War prompted

S. Bhalai, *The History of Mining and Geological Surveys in Jamaica*, Historical Geography and Geosciences,
https://doi.org/10.1007/978-3-031-42604-9_5

Britain's awareness of the responsibility for the development of its colonies, and the CD&W fund was established to pay for expansion of services in the poorer colonies (Wilson 1985).

The Commission's recommendations published in February 1940 revealed that social services in the Caribbean region were highly inadequate to meet the needs of the population, and none of the colonies could sufficiently provide for themselves in bringing into place the services to acceptable levels and encourage national development (Anonymous 1947). The Commission therefore recommended the establishment of a West Indian Welfare Fund financed by a yearly grant of £1 million from the UK Exchequer. This programme would be done for 20 years and administered by a special oversight body headed by a Comptroller who was guided by expert advisors. The money would be dedicated to schemes for improvement of education and health care, housing improvement and regularization, including slum clearance, establishing labour departments, land settlements, engineering, and geological survey activities, and for the provision of social welfare facilities (Anonymous 1947). A Directorate of Overseas Surveys was also established to make proper topographic maps for these regions to support the development activities (Wilson 1985).

At the same time, the British Government was independently considering a Development Finance Policy which had resulted in the enactment in June 1940 of the first Colonial Development and Welfare Act (Anonymous 1947). This Act provided £5 million from the UK Treasury for social and economic development in the colonial empire for ten years ending March 31, 1951. An additional £500,000 per year during the period would be allocated for research. The British Government, in recognizing that both welfare schemes had the same objectives, combined the programmes. The concept of a Comptroller of Development and Welfare in the West Indies was retained, and that officer and the various advisors were appointed in September 1940. The Comptroller was the head of the Development and Welfare Organization in the West Indies, the driving entity recommended by the Commission and established by the 1940 Act (Anonymous 1947). The Comptroller and his advisors conducted their fact-finding missions to the West Indian colonies between 1940 and 1942, and the period 1943 to 1944 was spent analysing the challenges facing the colonies. They recognized that the conditions imposed by Second World War would retard progress. The war had resulted in a shortage of raw materials and trained human resources to work in the West Indian colonies.

It was recognized that the first CD&W Act had limitations in the funding amount and procedures for approval. Further, it was identified too that development in the colonies may progress slower than expected. Therefore, a new CD&W Act was passed in 1945 which more than doubled the initial allotment to the schemes to £120 million for the ten-year period ending March 31, 1956 (Anonymous 1945). It also made the operations of the Act more flexible, such as the elimination of the annual vote of approval. The money could be drawn upon at any time within the period, subject to an annual maximum of £17.5 million for development purposes and £1 million for research. The Secretary of State Arthur Creech Jones in his address to the House of Commons in 1945, advised that the Colonial Governments had the responsibility to execute the schemes financed under the Act and had the primary responsibility for drawing up comprehensive development plans that would become known as the Ten-year Development Plan. He also detailed the funding support that each West Indian colony could expect (Anonymous 1947).

In a circular despatch of November 1945, the allocations to the West Indian colonies were presented: Barbados £800,000, Belize £600,000, Guyana £2.5 million, Leeward Islands £1.2 million, Trinidad £1.2 million, Windward Islands £1.85 million, and the other islands including the Bahamas £850,000. Jamaica got the largest share at £6.5 million. The Cayman Islands, and Turks and Caicos Islands at that time were dependencies of Jamaica. The sum to all the

British West Indian islands totalled £15.5 million for each year of the ten-year period in addition to that already spend and allocated up to March 1946 under the 1940 Act. The colonies would also benefit from a share of £23.5 million for central services for the colonial empire. These included research, higher education, topographical and geological surveys, and meteorological and aeronautical wireless services.

In his despatch, the Secretary of State requested that all the Colonial Governments should draw up development plans with the objective:

> . . . to ensure that all the resources available are used to the best advantage that the whole field of possible development and welfare is surveyed, and that the sum to be devoted to each project are determined so that the programmes form a well-balanced whole. (Anonymous 1947)

This compelled all the Governors, for the first time, to look comprehensively at the needs of their colonies and to put them into perspective. Like the recommendations of Governor Richards, the Secretary encouraged the colonies to regard the allocation as but one method of financing their development plan. He advised that local resources should be fully incorporated and utilized, and that each colony should estimate how much it could afford to acquire in loans and from increased taxation. The CD&W Act represented a realization that social advance was impossible without economic advance (Anonymous 1947).

5.3 Colonial Geological Survey

The Secretary of State for the Colonies, Creech Jones in his 1945 circular despatch, advised Jamaica of the grant of funds and mentioned that a scheme for the improvement of the geological survey of the Colony should also be considered (Anonymous 1946a). The Jamaican Government got started on their ten-year plan and it seems, inadvertently omitted this activity. Fred L. B. Evans, independent candidate for eastern Westmoreland, in early 1946, highlighted this, stating:

> I don't think many of us appreciated the value of bauxite mineral that we have in this country. And one day we got up and discovered that it was of tremendous importance to the Country. (Anonymous 1946a)

Later that year, in his Memorandum on Colonial Mining Policy (Anonymous 1946b), Creech Jones stressed the importance of the geological survey particularly as a primary object of the policy. He reminded Colonial Governments that provision was made under the CD&W Act for colonial geological surveys. To control the distribution of the geological survey allotments the Directorate of Colonial Geological Surveys was established (Wilson 1985). He advised that Frank Dixey was appointed as Geological Advisor to the Secretary of State for the Colonies. Dixey was formerly the Director of the Geological Survey in Nigeria. As the new Director his first assignment was to prepare a scheme for the staffing of the Colonial Geological Survey service (CGS) (Anonymous 1948a). Dixey, who originally was a lecturer at University College, Cardiff was enlisted in the military. He was released at the end of the Second World War and sent to Sierra Leone, as he puts it:

> . . . to a geologically unknown country to single-handedly conduct a mineral survey of it with no trained staff, field maps or any facilities for such activities. (Wilson 1985)

At the time that country had limited rail network and poor roads. The terrain was challenging but Dixey managed to successfully complete the surveys. For some of the areas he had mapped, later geologists declared them inaccessible.

The CGS had primary objective to identify mineral resources in the West Indies to guide future policy. Geological maps of all the colonies would be prepared to support the discovery of mineral deposits. The CGS began on January 1, 1947, at the Imperial Institute. Two years passed before the new organization took shape; however during this time Dixey travelled extensively, visiting existing geological survey organizations, and meeting with Governments of the territories where new surveys were required, but which had no local survey organizations.

It was originally planned that the CGS, based in England, would dispatch teams of surveyors to the colonies then return to produce the final printed maps at home. Dixey however, encouraged the colonies to establish or expand their own survey organization to provide general advice while the CGS would supply specialist services from its internal photogeology (geological mapping using aerial photographs) or geophysics expert, or consult with external services such as applied geochemistry from the Imperial College, palaeontology from the British Museum of Natural History, or mineralogy or chemistry services by the Imperial Institute (Wilson 1985). When the CGS was established, there were 11 Colonial Governments with geological surveys. By 1957 there were 13 new colonial survey agencies added, though some, such as that in Hong Kong, were temporary. With colonies becoming independent the CGS was later renamed Overseas Geological Surveys (Wilson 1985).

For his Caribbean mission, Dixey visited Jamaica, then on to other countries such as Trinidad, Belize, Guyana, Barbados, among others. He arrived first in January 1948 to determine the requirements for making a geological survey, and on his recommendation a team of geologists would be deployed to the island. He specified that the geological survey of Jamaica would lay the foundation for fully understanding and appreciating the local soil conditions and groundwater resources, and it would prove useful for engineering purposes and determining the location and size of mineral deposits particularly in the Blue Mountains (Anonymous 1948b). To accurately plot geological information, reliable topographic maps were needed, and fortunately at the time a series of twelve 1:50,000 scale topographic maps were being prepared by the Central Topographic Survey Organization. When Dixey made his assessment, six of these maps were already sent to Jamaica for revision and the other set would follow. Maps prepared from aerial photographs at 1:25,000 scale were also available, in addition to the original photographs used, which were taken in 1941. Dixey noted that these outputs of the topographic survey would greatly enhance the geological work and recommended geological mapping on the same scale as the new 1:50,000 topographic sheets. On staffing, Dixey recommended that provisions be made for two geologists; one to be appointed at the end of 1948 and the other a year later, and their service should continue until March 1956 (Anonymous 1948b).

Around the time of Dixey's visit, Jamaica's ten-year plan was still being prepared. The draft was approved by the Secretary of State for the Colonies and was to be submitted to the Legislature. This, however, was the second attempt to obtain the first tranche of development funding under the Act (Anonymous 1948c). Dixey's recommendations it seems, overtook the submission of the plan, and by the end of February 1948 it was announced that Jamaica was invited to ask Britain for a grant to undertake a proper geological survey of the island (Anonymous 1948b). This came on the advice of Dixey who recommended the survey activity for five years and a proportionate grant for two additional years to undertake the scheme he had worked out. He estimated a total cost of £58,133 where a small organization would be established to conduct basic surveys for seven years. Of this amount £52,413 would be recurrent expenditures and £5,720 for capital expenses. The Secretary of State also communicated that he was willing to view favourably an application for full financial assistance for the next five years and assistance on a progressively reducing scale for the Geological Survey organization to be established. During the two latter years the Government was expected to make some contribution to the recurrent expenditures of the scheme. The House was therefore invited to approve an application to the Secretary for this grant under the CD&W Act and the Government would also decide if the geological survey activity should continue as an expenditure on its books beyond the seventh year.

On March 4, 1948, the House approved the proposal for the establishment of a local organization to carry out basic geological survey for seven years (Anonymous 1948d). The schedule of expenditures was tabled to include capital expenditures of allowances for field equipment, survey and lab equipment, office furniture, two station wagons and the cost of a small building. The Institute of Jamaica's property in Downtown Kingston was selected to house the office and laboratory of this new organization. The decision was taken by the Secretary of the Institute, Phillip Sherlock, and the Director of Public Works that the funds allocated for the building should be 'spent on a building of lasting value to the colony'. As such, an extension was made to the rear of the old Museum building with the construction of a two-storey building. The upper floor would serve as offices and the lower rooms for the Institute. Until construction was completed, a miscellaneous storage basement room would be used to temporarily house the Survey when the first geologist assumed duty at the end of 1948. The Institute was long affiliated with scientific research including the geological surveys. The organization kept and maintained the most complete set of geological records and publications, not just for Jamaica, but also the West Indies at its Science Library of the Museum, and its West India Reference Library. It also had the rocks and minerals collection of Sawkins and Brown from the first official survey in the 1860s. The collection of Charles Matley, Government geologist from the Geological Survey of the 1920s however, was still to be received. Well-known palaeontologist Charles Taylor Trechmann had also planned to donate some thousands of specimens to the Institute's collection (Anonymous 1948b).

5.4 Staffing the Geological Survey Department

Within the British Empire there was a shortage of new graduate geologists from the universities, and the Second World War exacerbated the issue. Students of chemistry and physics were permitted to defer their national service until after graduation, but this was not the case for geology students. Further, a deficiency developed when the older geologists who had delayed retirement beyond their normal retirement age as a war service, decided to leave the service. A record recruitment drive for geologists started immediately after the war. This was triggered by demand from mining and petroleum companies, universities, museums, the Imperial Geological Survey, and the Colonial Office. The boom in geological survey activities at that time required trained personnel to be deployed all over the world.

In a circular despatch of November 22, 1947, to Jamaica, the Secretary of State listed some recommendations on appointing the new geologists (Anonymous 1948b). Their appointments were to be on permanent and pensionable terms with the understanding that they may be transferred elsewhere if the country no longer required their full-time service. Additionally, after four years their positions should be reviewed with considerations of the progress made. If all the work was done, then Jamaica may no longer require a full-time geologist. Dixey realized that the war-induced shortage of trained geologists from the colleges and universities would continue for several years and was doubtful that geologists would accept employment in Jamaica unless the salary and conditions of service such as allowances were comparable to those offered in other places. The war had ended and general recruitment to the Colonial Service resumed on June 1, 1945. By March 31, 1947, a total of 2,800 general appointments of both men and women were made through the Colonial Office Appointments Department out of 14,500 applicants. Many of the posts filled were connected to schemes approved under the CD&W Act 1945. The individuals selected were posted in 33 different territories in the colonial empire, with the largest numbers going to Africa and Far Eastern territories. The work force was deployed on a wide range of Government activities, including geologists and other related experts for colonial mining and geological survey departments. Many of the persons were from the UK but there were some persons selected from the Dominions and the colonies (Anonymous 1949a).

During the recruitment period, despite the large number of applicants, only a small number of persons were applying to serve as geologists at the CGS, and this shortage was evident. It was therefore in the best interest of Britain to publicize the need and encourage training of geologists to work throughout the colonies. The Secretary of State sought assistance from all the university appointment boards and committees in England, including recognized academic bodies (Anonymous 1949a). He kept them informed of the present and potential vacancies in the Service and sought their recommendations for suitable candidates. Dixey recognized the possibility that the incumbent geologist for Jamaica would likely be young and inexperienced. The salary and emoluments could easily be worked out to make an attractive package, but to ensure good quality work some oversight would be needed. He recommended that the survey should be undertaken with the guidance and general technical advice of an established Geological Survey organization such as that existing in Guyana. The Director or a Senior Officer from this survey should visit once or twice each year for advisory purposes. The Governor of Guyana and the Acting Comptroller for Development and Welfare agreed with this proposal. They further concurred that the Guyanese Geological Survey should be regarded as the central body where the West Indian colonies should seek advice and assistance as they may occasionally require (Anonymous 1948b).

This recommendation for the Guyanese Survey to be the main advising body was not well-received in Jamaica. Ivan Stewart Lloyd, Member of Parliament for Eastern St. Ann confidently proposed in March 1948 that the Jamaican Geological Survey instead should be entrusted with the responsibility for supervising geological work throughout the Caribbean.

> We should see to it that whatever institution or department or organization is set up to make geological survey in Jamaica, it should be entrusted with the responsibility of supervising geological work throughout all the Caribbean, that is, throughout all those units in the Caribbean we are attempting and hope to federate as one unit at an early date. (Anonymous 1948e)

The Federation he mentioned was an attempt to group the British Caribbean islands for administrative and other purposes (Black 1976). He went further to state that:

> . . . whatever organization is set out, [it] should be set up in conjunction with the West Indian university which is being constructed at present. I think it is inconceivable that we should go to the Jamaican Institute in order to spend that money to put up a building there and set up a department there and bring a skilled experienced geologist here without taking advantage of his services to train Jamaicans, Barbadians or Trinidadians who want to study geology in connection with the university we expect to set up here at an early date. (Anonymous 1948e)

This West Indian university was the University College of the West Indies which was established in Jamaica through a grant of £1.5 million from the CD&W funds to build and equip the institution. The site chosen was approximately 653 acres of the Mona Estate in St. Andrew leased from the Government of Jamaica for 999 years at a peppercorn rental. The annual maintenance cost was to be borne by the contributing colonies based on their respective populations. Seven colonies agreed to contribute to the upkeep; however, about 45% was covered by Jamaica. A teaching hospital was also constructed and equipped at £1 million from the CD&W fund. The university was not affiliated with the University of London but by special agreement the degrees awarded to graduates during the formative years were those of the University of London (Cover 1953).

The source for geologists to serve in Jamaica remained uncertain, but under the European Recovery Programme (ERP) there was provision for technical experts from the USA to be sent specifically for defining and securing raw materials for stockpiling. It was possible to obtain a geologist from the USA for the Jamaican Survey, and this idea was explored in late 1948. However, it was traditional that the appointments be made among British subjects, especially because the post was long-term. A request was made to the American Economic Cooperation Administration (AECA) to provide 50 persons to fill vacancies in colonial organizations conducting topographic, geodetic, and geological surveys. This invitation, however, was specific and not extended to British professionals, the British universities, the Royal Geographical Society, or any of the British air survey firms (Anonymous 1948f, 1949a).

In an early January 1949 sitting of the House of Commons, the Secretary of State for the Colonies, Creech Jones was questioned about this. He responded that the AECA was asked to supply 22 geologists and three chemists for short contracts not exceeding three years. These persons would assist in geological assessments in the colonies. These US officers would fill temporarily, the vacancies in the CGS which are open due to the war-induced shortage of suitably trained British nationals. He reassured that there was continued effort to obtain the services of UK and Dominion geologists for permanent service within the CGS and contact with these personnel is maintained through the Geology departments in the UK and Dominion universities, and through appointment boards. Specific vacancies were also advertised in the press and scientific journals (Anonymous 1949a).

The Secretary announced however, that exhaustive inquiry revealed that all the posts of the CGS departments would not be filled until about 1952. By that time the service would have expanded to 170 geologists from the pre-1939 figure of 50 geologists. He stressed that only first-class geologists were acceptable, and these were limited and insufficient to meet the demands of the British Empire. Creech Jones indicated that the officers from the USA would be paid by the US Government from the ERP funds in US currency and their local allowances would be paid from a special Sterling account. It was uncertain if any of the US officers would be deployed to Jamaica, though it seemed unlikely (Anonymous 1949a).

5.4.1 The Geologists

By the end of September 1949, it was announced that a two-storey eastern annex of the Institute of Jamaica would be constructed to replace the existing one-storey building. This would house the Survey staff, the bindery and additional storage for historical documents. This was one year later than the planned start-up date for construction. Dixey had modified his budget recommendation reducing the grant amount necessary to £41,450 from the CD&W funds. Of this, £3,000 was allocated for the expansion of the annex. The Secretary of State approved the grant and the Jamaican Government contributed £5,500 (Anonymous 1949b, c).

By mid-October, 45-year-old Professor Verners Aleksandrs Zans arrived as the first geologist to head the survey (Fig. 5.1). He was a Latvian refugee, from the capital Riga which was obliterated during the war. He had left in 1944 because of the political developments. Zans had studied in Riga, Stockholm, and Helsinki. He served as Chief Geologist (1937–1939) and Acting Director (1939–1941) of the Latvian Institute for the Exploration of Mineral Resources. In 1938, he organized a general geological survey in Latvia for purely scientific research and for economic minerals. Zans also lectured, serving as Associate Professor of Geology at Latvia University at Riga (1938–1944), and in a similar capacity at Baltic University at Hamburg (1944–1949) until his recruitment by the Colonial Geological Survey. Up to this time, he had authored 17 publications on geology (Anonymous 1961; Chubb et al. 1961).

A British Commission had selected Zans, along with other scientists to work for the British Government. Britain ideally sought to hire its own nationals for the Colonial Service, but due the shortage, the recruitment was extended to the refugee camps of persons displaced in Europe. Zans was interviewed in London during May 1949 and was given an offer to head the Survey in Jamaica. To prepare

Fig. 5.1 Technical staff of the third government-commissioned Geological Survey of Jamaica established in 1949. **a**—Verners Aleksandrs Zans (Director); **b**—Baron 'Barry' McGrath (Technical Assistant; first Jamaica employed); **c**—Howard Raymond Versey (Geologist); **d**—Lawrence John Chubb (Palaeontologist); **e**—John Bernard Williams (Geologist); and **f**—Edward Robinson (Geologist) (Courtesy of the Mines and Geology Division, Jamaica)

for the job he returned to Germany where he spent some months learning English. He returned to London in early September 1949 for a short training in photogeology at Imperial College, a specialized course conducted for geologists employed in the Colonial Service. It involved routine lectures and laboratory exercises along with visits to the Hunting Aero Surveys company to view the photographic equipment used on the aeroplanes. Hunting would later survey nearly the entire Jamaican landscape in 1953 and 1954, and those photographs would be used to support the geological mapping programme. The group also visited the offices of the CGS where they saw first-hand the production of geological maps from the aerial photographs. During the course, Zans met Lawrence John Chubb, Lecturer at University College who was planning his sabbatical in Jamaica for the first nine months of 1950 (Fig. 5.1). By October 4, 1949, Zans was on his way to Jamaica (Chubb et al. 1961).

Zans, accompanied by his wife Valentine and their two children Valdis and Ruth, arrived in Jamaica from the UK on the *S.S. Manistee* on October 18, 1949. The family was met at the pier by Captain Arthur Frederick Thelwell, Commissioner of Lands, in his capacity as Commissioner of Mines, as the Geological Survey was a part of the Lands Department

(Anonymous 1949d). Zans had also carried his own petrographic microscope, a *Carl Zeiss Jena*, which it is said, travelled in his children's pram when the family walked from Colditz in Saxony, to Hamburg during the war (Fig. 5.2) (Anonymous 1961; Chubb et al. 1961).

Zans was appointed to the Civil Service in Jamaica on October 18, 1949, and was given office at 70 Duke Street, Kingston (Anonymous 1949e, 1950a; Cover 1953). He received technical assistance from Baron McGrath (Fig. 5.1). McGrath, trained in Mining Engineering in Cuba, was the first Jamaican employed in the Survey as Technical Assistant, then Chief Surveyor. He oriented Zans in the culture and lifestyle of the colony, and further helped to on-board the other foreigners who would later join. He provided technical expertise as well as providing 'peace of mind' in the field. He often armed himself with a shotgun, more as a comfort for the expatriates who were cautious about the colonists. McGrath's would later distinguish himself in the immense contributions made in cave surveys (Chubb et al. 1961). Next to join the survey as geologist was Howard Raymond Versey who arrived from England late in 1949 and was appointed on December 31 (Fig. 5.1). The employment of only two geologists was planned but Chubb, whom Zans had met in London had arrived in Jamaica on sabbatical and he initially gave voluntary service (Fig. 5.1). By February 1950 all three geologists were engaged in planning the work programme for the geological survey of the entire island. Observing the need for personnel, Chubb, who was 62 years old, extended his time in Jamaica and accepted employment; being appointed on October 1, 1950. (Anonymous 1950a).

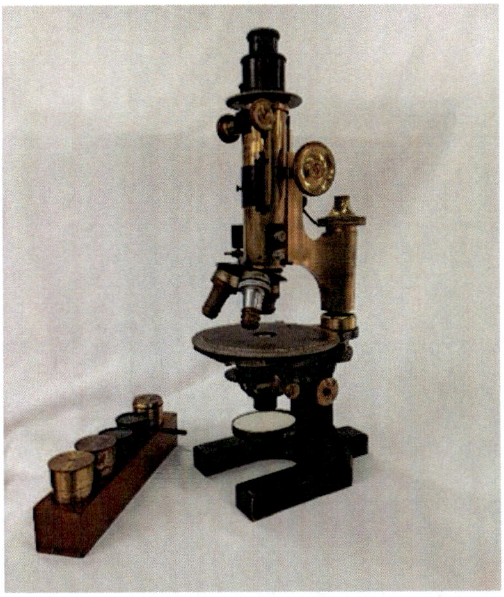

Fig. 5.2 *Carl Zeiss Jena* No. 35404 petrographic microscope carried by Director of Geological Surveys, Verners Zans to Jamaica. It is said that the instrument travelled in his children's pram when his family walked some 270 miles from Colditz in Saxony, to Hamburg during the Second World War (Chubb et al. 1961). It stands at 14 inches and is complete with oculars of varying magnifications, along with accessory plates. It is currently kept at the Mines and Geology Division (Photo by Author)

The Geological Survey was formally established in October 1949, and by 1950 it was fully staffed with surveyors, draftsmen, and field assistants. Considerable work had already been undertaken and the basic geology of the island became available in summary, and oftentimes reproduced in the annual 'Handbooks of Jamaica' (e.g., Cover 1951), a thick report on the features of Jamaica, the economy, the people, and the operations of the various Government organizations among other things. The Geological Survey Department of ten staff members including the three geologists was attached as a Branch of the Lands Department of the Ministry of Mines and Lands. Temporary office space was given at Duke Street with the plan that the Survey would become a separate department when it occupied the new offices at 14 East Street. Construction of the new building at the Institute, however, was delayed, and at other times progressed slowly. The cost for the new office had also exceeded the original budget. In August 1950 the office was six weeks to completion when it was reported that the approved vote of £8,500 for construction had been exhausted. The Government therefore made a proposal for a special grant to complete the work. By early March 1951 it was announced that a three-storey building in the eastern courtyard had been completed and the Geological

Survey would occupy the third floor (Fig. 5.3). The new building had to its rear, Georges Lane, and hosted offices for the Institute of Jamaica, including the museum and lecture gallery (Anonymous 1950b, c, 1951a; Cover 1953).

Shortly after the Geological Survey moved to its own space, it was established as the independent Geological Survey Department of the Ministry of Agriculture and Lands effective April 1, 1951. The new department had the primary function to map the country geologically on a scale of 1:50,000, to conform with the twelve topographic sheets prepared by the Directorate of Colonial Surveys as recommended by Dixey, Geological Advisor to the Colonial Office and Director of Colonial Geological Service (Cover 1953; Anonymous 1948b). The survey was also mandated to investigate the mineral resources of the island. The geological maps produced would serve as a basis for water supply investigation and soil mapping, and would provide information for engineering works such as suitable sites for dams, and bridge foundations. Additionally,

Fig. 5.3 Staff of the new Jamaican Geological Survey Department outside the East Street, Kingston office in the early 1950s. *Front row* (l – r) Howard R. Versey (hand on wall), Frank Dixey, Director of the Colonial Geological Survey, Verners A. Zans (one step above), Lawrence J. Chubb, Unknown. *Middle row* (l – r) Winston Jackson (back to column), Mike B Cunningham Scott (finance and accounting), Stanislaw Vincenz (geophysicist with the Industrial Development Corporation; part-time at the survey), Basil Bailey; *Back row* (l – r) Clive Adams, Baron McGrath (Courtesy of the Mines and Geology Division, Jamaica)

sources of raw materials for various industries such as for making cement, bricks, gypsum boards, and pottery would be identified. The mineral potential of the Blue Mountains and similar areas was surveyed in detail, but priority was given to identifying water supplies in drought-stricken areas. All these roles were deemed a necessity for the Survey to undertake based on the development plan for the colony. The department occasionally carried out special investigations on behalf of other departments or private firms, and sometimes for the latter, a small fee was charged. Following the complete survey of each topographic sheet a geological map and corresponding memoir were published. It was decided that on completion of the surveys a final report on the geology, water supply and mineral resources of the whole island would be compiled (Cover 1953).

5.5 The Early Years

Three years into operation of the Survey in November 1952 an application was made to the Secretary of State for the Colonies for a further grant of £28,300 from the CD&W funds to a total expenditure of £48,300 to accelerate the geological survey programme. In an address to the House of Representatives, it was announced that the surveying activity, though approved from 1948, had started in 1949, and despite being scheduled for seven years, would not be completed in less than 11 years unless the number of geologists is increased. The proposal was made to increase the staff to five geologists to expedite the work to completion by March 1958. The extra funds would come from the Government, loans, and general revenue over a three-year period. It was also recommended that additional staff should be sought from the United Nations which had a Technical Assistance Programme (Anonymous 1952a, b).

In November 1952, a quarter of the Jamaican landscape had been mapped and this rate of output that was used to deduce the 11-year duration to complete the surveys and prepare the maps and supporting booklets using the current staffing. The proposed staff increase would enable completion in five and a half years, that is, by March 1958 (Cover 1953; Anonymous 1952b). The spending was therefore justified that:

> This is done in view of the current economic conditions of the country which require that all development schemes, agricultural, industrial, and mining should be accelerated, since most of them depend on the maximum results of a preliminary geological survey. It is essential that the programme should be completed as rapidly as possible. (Anonymous 1952b)

In 1952 the United Nations was approached for a micropalaeontologist and a hydrogeologist. It was in May 1956 that Glenn C. Prescott, geologist of the United States Geological Survey (USGS) arrived in Jamaica under the Technical Assistance Programme. He was tasked to initiate a Ground Water Branch and to train a geologist in the Geological Survey Department in techniques and methods of studying and analysing groundwater problems. Versey was attached to the new branch with the intentions to become the head at the end of the training period (Government of Jamaica 1957). Arrangements were also made with the Industrial Development Corporation (IDC) for the full-time engagement of a geophysicist using the amount budgeted in the Survey's allocation for geophysical work. That scientist would be made available to undertake the work of the Survey when needed. The IDC was established in 1952 to promote industrial growth through new factories, engineering services, industrial training and encouraging investment in industry. It opened the first industrial estate and made industry the largest contributor to the gross domestic product of the island. The corporation, in selecting suitable industrial sites conducted engineering surveys; therefore, having a permanent geophysicist was of great value. Stanislaw A. Vincenz was therefore appointed in the mid-1950s as the Government geophysicist (Anonymous 1957a). The Survey's work was also supported by the employment of two new geologists, John Bernard Williams, appointed late November 1953, and Edward Robinson who was appointed early November 1956 (Fig. 5.1) (Government Public Relations Office 1961).

The monitoring plan for the Geological Survey Department included an annual visit by the Director of the British Guiana Geological Survey. The CGS had arranged that the Director in Guyana would be an Advisor to the all the Geological Survey organizations in the British Caribbean. Therefore, when the Jamaican Survey was established, Smith Bracewell was the head in Guyana, and as part of the regular working relationship, Zans was required to consult with him, as well as the Director of the CGS. Bracewell was succeeded by E. R. Pollard who, in March 1953 gave way to Gordon Stockley who had worked in Jamaica in the 1920s as Government Geologist and had just demitted office as Director of the Geological Survey of Tanganyika (Anonymous 1954a; Beard 1954). Stockley spent three weeks in Jamaica in January 1954 where he did much travelling with the geologists examining developments in the bauxite, gypsum and cement industries, and the iron prospects. His duties as Advisor required him to spend two months each year travelling, and he had come to Jamaica with his wife Daisey, whom he had met and wed here back in the 1920s. Immediately before coming to Jamaica he had spent three weeks in Belize, and before, in May and June 1953, he visited the Windward Islands. Before, in 1949, Stockley had returned to Jamaica with his wife to conduct hydrogeological surveys on the estates belonging to West Indies Sugar Company (Anonymous 1954a, 1954b).

5.5.1 The Work Programme

The Geological Survey Department was given the mandate:

> . . . to carry out a complete systematic survey of the entire island, its mineral resources and underground water supply potentialities. (Government of Jamaica 1956)

The organization conducted geological mapping at the 1:50,000 scale and the information compiled to prepare an island-wide *Provisional Structural and Geological Map of Jamaica* in 1952 (Zans 1952a). This map was oftentimes reproduced in the Handbooks of Jamaica in the 1950s (Zans 1953). By the late 1950s the information was refined and a 1:250,000 scale geological map of the island dated 1958 was prepared, probably to coincide with the expected end date of the survey (Fig. 5.4; Zans 1958). The map was sent to the Directorate of Overseas Geological Surveys (formerly the Colonial Geological Surveys) and was published in 1959 (Crooks 1987). An explanatory booklet or synopsis (Zans et al. 1963) was prepared later to support the map. The Government, in seeing the value of the Survey's findings, made the organization a permanent establishment (Anonymous 1954c).

The core focus of investigating the mineral wealth of the island led the Survey to publish its first Bulletin in 1951 entitled *Economic Geology and Mineral Resources of Jamaica* (Zans 1951a). It represented the comprehensive findings of two years (1949–1951) of work along with that from the earlier geological survey programmes. It generated enormous interest from both local and overseas industrialists who found it a reliable reference for the available mineral resources in the colony. The Director stated that 'it should give a brief and comprehensive account of the present knowledge of the economic mineral resources and potentialities of Jamaica'.

When the new manual was released, it was declared that:

> All in all, it provides a modern background to Jamaica's mineral resources which might well be the foundation textbook of a new mining age. (Anonymous 1952c)

It propelled an increase in the number of applications for prospecting licences, particularly for the new and attractive mineral deposits identified. This included deposits of silica sand, kaolinite, and iron. At Hodges, St. Elizabeth, a high-grade silica sand associated with kaolinitic clay was discovered. The Geological Survey had estimated that between 500,000 to 1 million tonnes of quartz sand was available. The sand was pure silica which made it suitable for making glass.

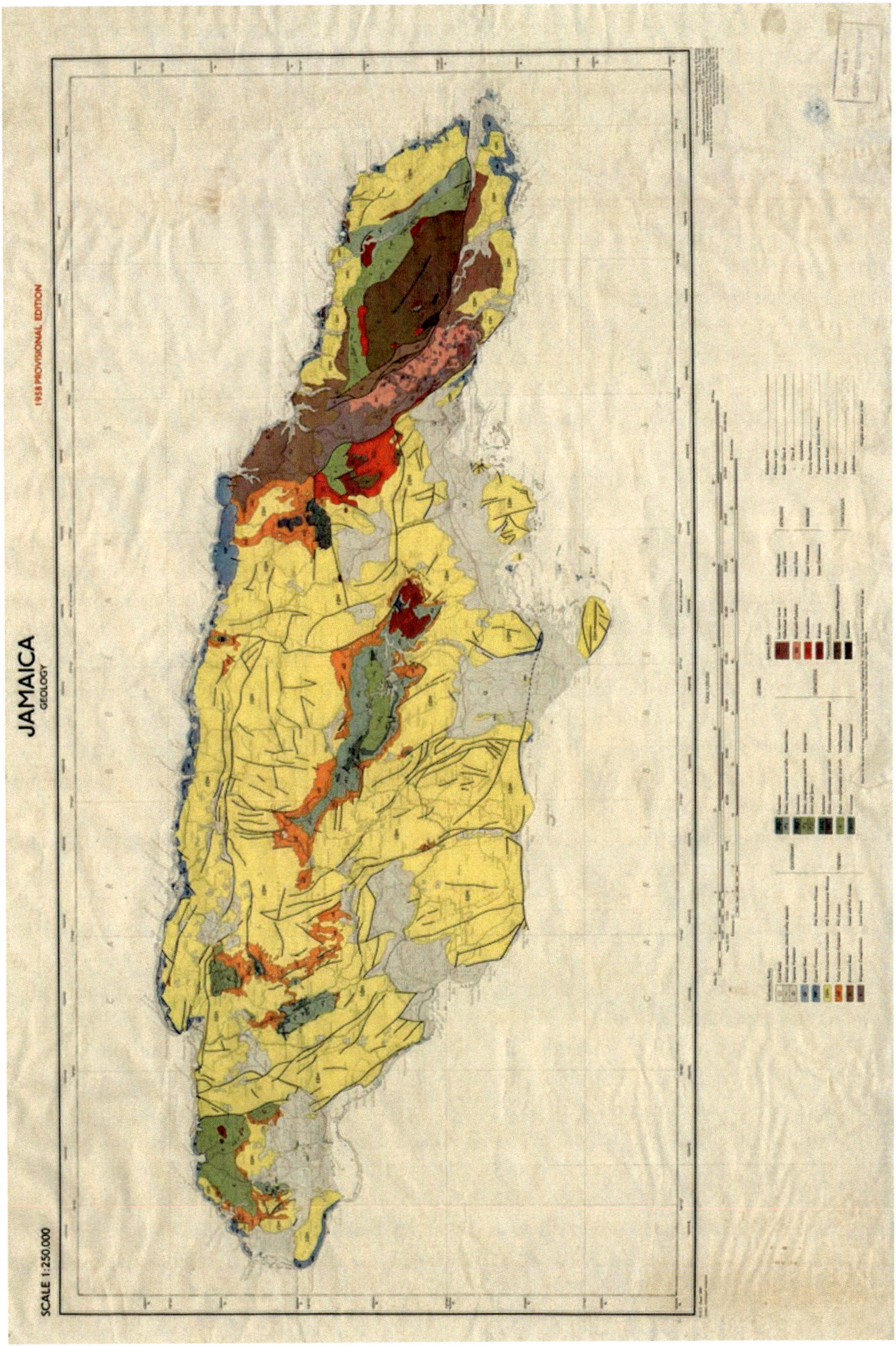

Fig. 5.4 Geological map of Jamaica (1958) prepared at the 1:250,000 scale. This map was released in 1958, likely to coincide with the scheduled end date of the survey (Zans 1958) (Courtesy of the Mines and Geology Division, Jamaica)

This interested the Caribbean Chemical Company Ltd to consider it for the manufacture of bottles (Anonymous 1955a). At Glade Orchard, St. Andrew, magnetite and hematite bodies averaging 20 ft. thick with assays of 64% iron estimated at over 1 million tons were discovered in February 1951. In 1953 and 1955 magnetic surveys were conducted; the latter by Aeromagnetic Surveys of Canada supervised by Vincenz, the geophysicist from the IDC. Local companies, Jamaica Iron Ore Limited, and Mavis Bank Iron Ore Limited were formed to develop these deposits. The magnetic surveys were extended to other areas such as northern St. Catherine, along the northern margin of the Benbow Inlier where a distinct anomaly, possibly another iron lode, was identified (Government of Jamaica 1958a).

To enhance the potential for discovery success for metallic minerals a regional geochemical survey of the major streams on the eastern side of the island was undertaken in 1960. The Blue Mountain area was deemed to be a rich metalliferous province as suggested by the abundance of old Spanish mine workings, especially for copper and gold. In addition, the old mines in Clarendon parish were explored but overgrown vegetation restricted access and some shafts were never located. The metallic mineral surveys also included an assessment of the gallium content in Jamaica's bauxite. The potential for gallium was known from earlier research on bauxite, and in 1961 samples were sent to the Overseas Mineral Resources Unit in London for quantifying the gallium. The importance of gallium in semi-conductors had increased in the 1960s (Government of Jamaica 1961).

In addition to the metallic minerals, the Survey explored the varieties of marble on the southern slopes of the Blue Mountain in 1961. The Mount Hibernia and Serge Island deposits in St. Thomas parish were surveyed in detail, and its economic potential determined. The following year the Serge Island deposit was brought into economic production by Serge Island Jamaica Ltd. Tiles from that facility were used in the decoration of many prominent local buildings. Other materials studied include the clay deposits of Westmoreland parish that were quantified in 1965 to reveal approximately 250 million tonnes suitable for structural ceramic. In the same year 150 million tonnes of dolomite were identified at Port Henderson (Government of Jamaica 1963, 1966).

The Geological Survey also advised on engineering geology matters. Assessments were done on several development projects such as large construction sites, dams, and airports. The Jamaica Broadcasting Corporation, for example, commissioned a survey in May 1953 of the reef off the coast of Ironshore, St. James to determine the suitability for a wireless mast for public broadcasting. In 1954, power provider, the Jamaica Public Service Company requested an assessment for the boring of a 2,200-foot-long tunnel for the new Rio Bueno hydroelectric power station. The Kaiser Bauxite Company, as part of their Port Rhoades development needed to cut an underwater channel and as such also hired the Survey to conduct assessments of the seafloor in the harbour at Discovery Bay, St. Ann to make it accessible to large ore carriers (Cover 1953; Government of Jamaica 1955, 1961). At Point, Hanover, ground conditions along the coastline were investigated in 1961 for the construction of a sugar and molasses wharf. In that year too heavy rains associated with a hurricane in October triggered a large landslide at Money Hole Corner, along the Junction Road from Kingston to St. Mary. The Survey identified that the rocks were weakened due to crushing, and this was the manifestation of the major Wag Water Fault. The locality got its name as the residents oftentimes obtained on-the-spot employment in clearing debris from the road when the hillside failed. Soil investigations were also done on 150 acres of reclaimed land at Foreshore Road in Kingston in 1963 for a project called Newport West. The plan was to reclaim a total of 300 acres from the sea to a designed eight feet above sea level and extending 600 ft. from the existing shoreline (Government of Jamaica 1964, 1965).

5.6 Oil and Gas Exploration

The Survey was also responsible for investigating oil and gas occurrences. One peculiar case was oil observed at Moore Town, Portland, in 1959. The oil was tested and found to be refined oil, possibly kerosene. Trenching along the trace of the seepage revealed a spillage on the bank of the Negro River (Government of Jamaica 1962).

Late in 1951 the Jamaica Government issued a licence for oil exploration to Base Metal Mining Corporation of Toronto. Oil exploration was new in Jamaica, and Zans was careful to explain the procedure to many curious Jamaicans. He listed the steps used by modern companies and explained that commercial quantities of oil were not expected near the surface but likely at several thousand feet depth (Zans 1951b). Base Metals contracted Lundberg Explorations Ltd of Toronto to carry out airborne radiation geophysical survey of 2,800 square miles or nearly two-thirds of the island's surface. Four radiation patterns, presumed to be caused by subsurface features such as oil and gas pools, were identified at Cambridge, St. James; Yallahs, St. Thomas; northeast of May Pen, Clarendon; and 2 miles south of St. Ann's Bay near to Windsor where there is a known natural gas seep at the surface (Anonymous 1955b).

By mid-1952 an exclusive licence to prospect for oil and gas across the whole island was granted and arrangements were made for the first test boring. Zans explained that during the recent geological surveys, and even before, geologists had found potential sources of oil such as bituminous limestone at Thorton in St. Elizabeth, and at Cambridge. However, considering the conditions of neighbouring Haiti and Cuba he did not expect large quantities but would still await the findings of the first borehole. The Geological Survey worked closely with the exploration company and during the first stage it provided geological information including identifying sites where deep-hole drilling can be done. The only deep hole ever drilled in Jamaica was by Jamaica Bauxites Ltd in a topographic depression half-a-mile south of the Kendal Railway Station, Manchester, in 1945. This hole, a water-well had extended to 1,574 ft. and penetrated layers of lignite, tar sand, bituminous shale, and limestone between 1378- and 1468-ft. depths. An oil well on the other hand extended several thousand feet deep (Anonymous 1955b).

Oil exploration was previously done in Cuba, Haiti, and Dominican Republic. In Cuba, the deepest exploration well penetrated more than 11,000 ft. At the time Base Metals started their hole the general success rate in exploration was one productive borehole in each nine or ten drilled (Anonymous 1955b). Drilling did not start until mid-1955. That boring was made at Negril Spots, Westmoreland. On June 21 drilling stopped as it was announced that the 'stratification of the earth' was promising of oil. Between 2,510 and 3,430 ft., porous dolomite, a potential reservoir rock was encountered. Being an exploration company and not an operating company, Base Metals sought the advice of Esso Standard Oil, S.A. The head of Esso's Exploration Division and the company geologist visited the island. The Negril Spots hole was drilled to 6,314 ft. (Anonymous 1955c; Wright 2010).

Esso decided to pursue and Stanolind Oil and Gas Company, a wholly owned subsidiary of Standard Oil of Indiana (Stanolind), one of the five largest companies in the USA, acquired major interests in Base Metals Corporation oil rights. The company declared that US $2 million would be expended on geological and geophysical surveys, and drilling of holes. The announcement was made by Jamaica's Chief Minister, the Honourable Norman Manley who negotiated the agreement on behalf of the Government in October 1955. He declared the acquisition of the oil rights and the finalization of a new oil exploration and exploitation agreement, which would be signed simultaneously the morning of October 26, 1955, in Kingston, New York, and Oklahoma (Anonymous 1955d, 1955e).

Stanolind, through their subsidiary, Pan Jamaican, started drilling in 1956 in the Santa Cruz Mountain, 5 miles south of Malvern where traces of gas were observed (Wright 2010). Later, the company drilled holes at Cockpit, Trelawny, then started a hole in west Negril. There was no showing of oil or gas, and in September 1957, the company announced that it would stop boring noting that the hole was 'dry and abandoned'. By November 1957 the company announced that it would abandon exploration in Jamaica. It had spent US $4 million in 20 months of exploration. Their licence would expire in June 1958, but the company had already expended double the budget and drilled three wells without success (Anonymous 1957b; c).

The unsuccessful on-land exploration was discouraging, however, the Geological Survey reported to the Government that the offshore Pedro Cays, an area unexplored by Pan Jamaican, were most likely a favourable target (Anonymous 1957d). In the mid-1960s the Commissioner of Mines collaborated with the Director of Geological Surveys, Howard Versey in proposing that the Government approach the Canadian External Aid Office to request an expert geologist specializing in problems of petroleum prospecting to review the existing prospecting work and geological knowledge. Hugo R. Greiner of University of New Brunswick was selected. He had vast experience in the oil and gas fields of Canada and was also an expert in palaeontology, stratigraphy, and sedimentology. As a result, *The Oil and Gas Potential of Jamaica* was published in 1965 as the fifth bulletin of the Geological Survey Department (Greiner 1965). Greiner supported the suggestion of the Survey to explore Pedro Banks. Thirteen years later oil exploration resumed, where offshore at Pedro Banks, and on land at Portland Ridge, Clarendon were drilled; both with no success (Wright 2010).

5.7 Promoting Geology

At a June 1951 lecture at the Institute of Jamaica, while speaking to the Natural History Society about the Geological Survey, Lawrence Chubb explained that Jamaica has many geological problems to solve. He however highlighted that the first problem to overcome was not geological, but in getting the people of Jamaica to understand the purpose of a geological survey (Anonymous 1951b). Since the establishment of the first Government-commissioned Geological Survey in Jamaica, the Government Geologists sought to educate persons on the rocks and various geological features of the Island. Lucas Barrett and James Sawkins kept specimens of rocks, minerals, and fossils at their office for educational purposes, with plans to establish a geological museum at Spanish Town. (Sawkins 1869). This was never fulfilled but the specimens they collected along with those gathered during the surveys were handed over to the Institute of Jamaica. The Institute, about 30 years later, established a new museum in 1895 where geological and natural history collections were maintained (Anonymous 1941). Unfortunately, it was reported in January 1940 that the collection of rocks originally collected and archived by the surveyors in the 1860s, was examined and found in very bad condition, as many had lost their labels and become useless. This was largely due to the 1907 earthquake and also general neglect (Anonymous 1940; Brown and Langner 2002). However, efforts were made to catalogue the specimens and by October 1941, the Institute's Secretary, Sherlock reported that the Sawkins' collection was restored and 'provided the nucleus for the museum' (Anonymous 1941).

Government Geologist, Charles Matley, in the 1920s, popularized geology and educated the public of the progress of the various activities and findings of his Geological Survey through his many extensive articles in the local newspaper *Gleaner* (Anonymous 1923, 1924; Matley 1923, 1924a, b). This was limited to those who could and cared to read, or even those who could afford to purchase the newspaper. The modern survey of the 1950s however took greater steps towards educating the citizens through presentations and exhibitions, donating collections of specimens to schools, founding a local geological society, establishing a Geology

Museum, and staffing and equipping the Geology Department at the University College of the West Indies.

At the June 1951 lecture to the Natural History Society, in speaking on the pervasive ignorance in Jamaica, Chubb noted that a great deal of apprehension existed, not only with the rural people, the farmers and labourers, but also among the Kingston businessmen. He stated that the usual question by the citizens is, 'What are you looking for?' and many believe the search was for a definite mineral such as metallic ore or oil. He stated however that the businesspeople did not ask questions but instead wrote newspaper articles. He explained that they too thought the Survey existed to find valuable minerals but revealed their unfamiliarity in their printed suggestions on how to achieve discovery success, such as those shared in early 1950s when oil exploration was about to start (Anonymous 1951b; Zans 1951b).

Chubb, retired faculty member of the University College London, promoted interest in geology, and it is certain that he got the support of Zans, who before the war, also lectured the subject. The efforts initially came through a campaign of radio broadcasts, newspaper articles, and participation in exhibitions and lectures at various schools and colleges (Anonymous 1956). In 1955 the Tercentenary Diamond Jubilee, or 300 years since colonization by Britain, was celebrated. The festivity was styled *Jamaica 300* and featured many commemorative activities which included publishing the *Jamaica 300 Supplement* of the *Gleaner* (Zans 1955). In this supplement Zans wrote on geology and mining in Jamaica. His article presented the successes and failures in the Mining Sector over the 300-year period. The celebration also involved a two-day agricultural show in the summer at Denbigh, Clarendon, where the Survey showcased a large geological map of the island measuring 16 ft. by 6 ft. marked with the different rock formations and the location of mineral deposits. There was also a relief model of Jamaica at half the size of the geology map, displaying drainage and irrigation, and topography. Over 100 local mineral specimens, rocks, fossils, along with literature and publications on Jamaican geology were also displayed. It was reported that, 'the display enjoyed wide popularity and attracted thousands daily, thus advertising the mineral resources and giving the Jamaican public a better understanding of the Department's work and aims' (Beard 1957).

Several Tercentenary lectures were organized, such as that held in May 1955 at the Wolmers Boys School to the Association of Assistant Masters and Mistresses (Anonymous 1955f). Chubb explained the value of introducing geology in the school curriculum to interest the local population in the subject. Chubb recommended the GCE Ordinary Level Geology as a good school subject where the exam questions traditionally set on the geology of England could be substituted with questions on local geological features. The lecture was subsequently printed as a feature in the *Daily Gleaner* and was broadcast on *Radio Jamaica*.

Chubb, a few days later, announced the proposal to form a 'Jamaican Group' of the Geologists' Association (of London) (Beard 1957). This association, based in London, was founded in 1858 to promote research in geology. Chubb noted that the membership at that time was 2,500 persons with less than 500 being professional geologists and the majority being businessmen, civil servants, school masters and mistresses, university teachers and students, and sixth form boys and girls. He mentioned that some geology hobbyists had pursued to the extent where their research became of real value. He explained the steps to becoming a member and the benefits which included access to meetings, field trips and the published *Proceedings*, the quarterly journal of the association. At the time only Chubb and John Williams of the Survey were members, along with a Mrs. S. M. Evans of Wolmers Boys.

By the end of July 1955 Chubb proudly reported that at the July 1 meeting in London of the Geologists' Association, 17 members were elected (Chubb 1955a). Among these were the Chief Minister and Minister of Agriculture, Norman W. Manley, D. A. Bryn-Davies of Alumina Jamaica, Robert Innes of the Sugar Research Department who was one of the scientists who identified Jamaican bauxite, Professor Zans and his Survey staff, and many others. Thirteen other members were also nominated for election in the November meeting. On Friday, October 7, 1955, an inaugural meeting, and formal dinner at Kingston's Liguanea Terrace Hotel marked the formation of the Jamaica Group; the first group to be formed outside of the UK. This hotel was owned and operated by Walter Butz and his wife. Butz was an American businessman and amateur geologist who became involved in mining Jamaican cave phosphates and the manufacture of wall boards from Jamaican gypsum. He was also a founding member of the Jamaica Group. The meeting was chaired by the first president, Chubb who expressed the satisfaction of the London Association which had sent a letter of encouragement from its Council, and a cable to welcome the formation of this first overseas group (Government of Jamaica 1958b). Zans announced that the Jamaica Group had grown to 62 members, though initially it was conceived to start with between 15 and 20 persons. He also encouraged amateur geologists, noting the important contributions they had made in the past. He further stated that the Survey …

> . . . is handicapped by the terms of reference which results in the Department devoting much of its energies to practical matters such as mineral exploration and water supply problems. The amateur geologist is under no such restriction and can pursue any line of enquiry that appeals. (Anonymous 1955g)

He reminded the group that in 1857 it was a keen amateur who examined a consignment of ore from the Charing Cross mine at the wharf in Kingston awaiting shipment and found that it contained gold (Anonymous 1955g).

The Jamaica Group convened its first annual general meeting in December 1956 and the local membership had increased from 62 to 100. The efforts and passion of Chubb the founder was credited for the increase. Chubb, however, acknowledged the contributions of his colleagues at the Survey and the management and technical staff of the bauxite companies. At the time, seven field excursions were scheduled along with four lectures and a radio address. The association also erected a tablet on the wall of the Halse Hall Great House where Henry De la Beche lived to commemorate his geological survey work in Jamaica and the UK. The tablet was unveiled by the Governor of Jamaica, Sir Kenneth Blackburne in December 1958 (*see* Fig. 1.5 *of* Chap. 1). Matley, on his visit to Jamaica in 1939, had originally proposed this idea but the effort was distracted by the war (Anonymous 1956).

Strong advocacy for the teaching of geology in schools was made at the December 1956 meeting. It was highlighted that geology studies were particularly necessary for Jamaica's development and that citizens were gradually realizing that the knowledge of the geologist was contributing to the increase in employment and wealth of the population, citing bauxite discovery and mining. It was announced that the Chief Education Officer had invited a science editor from England, and this was considered an indication that the Education Ministry was interested in developing science teaching in Jamaica. Chubb announced the intention to develop a small manual introducing geology, but he would use illustrations from Jamaica and nearby countries. By 1957 the teaching of geology began to be successfully incorporated in school curricula. Various high schools like Ruseas High (Hanover), Knox College (Manchester) and St. Georges (Kingston) taught geology, and the students began sitting the GCE Ordinary Level exam for the subject. Field classes for schoolteachers and sixth form students of Geology were conducted with Clarendon College (Chapelton), Ruseas, Knox, Camperdown College (Kingston) and St. George's College. Later, St. Georges College and York Castle High both received rock collections as teaching aid from the Survey in 1959 (Anonymous 1956, 1957e).

The Jamaica Group later became independent and evolved into the Geological Society of Jamaica. Initially, the group reported its activities including geological field trips in *Natural History Notes*, the journal of the Natural History

Society of Jamaica. However, in 1958 it began published its own proceedings as *Geonotes* (Robinson 1996), which evolved into the *Journal of the Geological Society of Jamaica* and later, the *Caribbean Journal of Earth Science* reflecting the broadened membership base and interest of the other Caribbean islands. (Robinson and Gentles 1980).

5.8 Geology Training at the University

Having successfully incorporated the teaching of geology at the high school level, the Jamaica Group at their December 1957 annual general meeting contemplated expanding the teaching of geology to the Tertiary level. It was stated that for years there were already repeated calls for a faculty in Geography to be established at the University College of the West Indies. The Ministry of Education had recognized that individuals desirous of teaching Geography had to seek qualification overseas as there was no local institution that offered this training. A dedicated training programme in the earth sciences would help to meet the demand for professionals both locally and globally. There was already an increasing interest in Geology as a subject in some Jamaican high schools. The Geological Survey Department assisted in practical classes, examinations, and field work to these schools, some of which taught Geology to the advanced level. Further, there were increasing developments in petroleum exploration in Trinidad (Anonymous 1957e).

The University College of the West Indies therefore, in 1959, organized a committee to establish a Department of Geology, and by 1960, approval was given to staff the new department. This committee included Lawrence Chubb, President of the Jamaica Group, who had vast experience in academia, having taught at University College, London, before coming to Jamaica. He had just been awarded early in 1958, a Doctor of Science (D.Sc.) degree by the University of London, the highest scientific degree obtainable. This was bestowed for published work of such quality that it gave the writer an authoritative standing in the subject and in his specific line of research. Chubb had received his Doctor of Philosophy (Ph.D.) in 1931 for work on the geology of the Pacific islands which he visited between 1921 and 1925 (Anonymous 1958).

At the new department, Geology would be offered as one subject in the University of London B.Sc. (General) degree. This meant that students, in reading for their degree, could major in Geology. At that time, the University College of the West Indies was in a 'scheme of special relationship' with the University of London where degree examinations were conducted by examiners from the University of London in association with examiners nominated by the University College of the West Indies. The degrees awarded were from the University of London. Under guidance from Professor T. H. Elkin, Geography was offered as a separate subject in the Geology Department later in 1965. The degree was awarded by the University of the West Indies (new name) since October 1962 when University College became an independent degree-granting university (Jamaica Information Service 1965; Robinson 2012).

The Geology Department, financed by the CD&W funds, became official in August 1961 and started with four staff members who occupied a dilapidated wooden building. This structure was previously occupied by the Chemistry and later, the Chemistry Technology Departments. The latter was relocated to the new Engineering Faculty at St. Augustine, Trinidad. Edward Robinson, who had started at the Geological Survey resigned as geologist in August 1961 to take up his post as lecturer in the new department. During that summer, the Geological Survey had just relocated from East Street to a newly constructed building at Hope Gardens to share offices with the Department of Mines which also relocated from their Half Way Tree offices. Kevin Burke the head of the new Geology Department, joined later in September as he was in London developing the syllabus and ordering books and equipment. Not surprisingly, Burke was Chubb's student from University

College in London. Sheila Versey was the part-time secretary until Pauline Lyn came on full-time. A Trinidadian, Mr Hamilton served as the rock-preparation technician. He also taught geological structures and built relief models of the Jamaican landscape.

Classes began in October 1961 with 26 students. Due to the lack of space, teaching was done in the Chemistry and Physics Departments. Classes were also held on Saturday mornings. Generous gifts of teaching specimen to support interactive laboratory work were received from the Geological Survey Department of Jamaica, the British Museum (Natural History), the Geological Museum of London, and individuals such as Lawrence Chubb and Howard Versey. The Geological Survey also handed over teaching materials they had used at the high schools, as the new Geology Department took over that role. Between 1958 and 1959 the Survey had received donations of specimens and materials from various members of the Geologists' Association for their use in teaching practical Geology (Williams 1962; Government of Jamaica 1963; Jamaica Information Service 1965; Brown and Langner 2002; Robinson 2012).

5.9 Water Resources Authority

Before 1954 there was no organized programme for the systematic collection of streamflow data in Jamaica. Some entities, however, collected this data for special purposes and for limited periods, but these were not intended for public use. These organizations included the Jamaica Public Service Company, the Kingston and St. Andrew Water Commission, and the Public Works Department. The Water Commission was responsible for the maintenance of large water supplies including distribution, along with the maintenance of the sewage system in Kingston. The Public Works Department had a Hydraulics Branch, and the Home Affairs Ministry which later became the Local Government Ministry had a Water Supplies Branch in charge of surface water. The Geological Survey had been tasked to explore reliable sources of groundwater, and this was done through its Groundwater and Drilling Branches.

Arising from a request by the Jamaican Government in 1952, the United Nations Technical Assistance Administration assigned specialists, initially from the US Army Corps of Engineers, and later from the USGS to undertake reconnaissance surveys to determine the needs of the island for water resources investigations. The specialists were charged to design projects for both surface and groundwater investigations, including the training of Jamaican personnel. This started in November 1953, and by February 1954 the first project for implementation was underway. The surface water assessment recommended the construction of 25 stream-gauging stations across the island along with the training of persons to operate the instruments, compute and compile the data, and publish the records. Within one year, 24 stations were constructed under the direction of hydrometeorologist, C. C. Yonker of the USGS. The Water Resources Section in the Public Works Department was established to execute the work programme connected to the establishment of the measuring stations. The Water Supplies Branch was responsible for the systematic collection of the streamflow data through gauging stations. For groundwater investigations, Glen Prescott, a geologist from the USGS, was deployed in May 1956 to conduct groundwater investigations in the Clarendon Plain of southern Jamaica, and to train personnel in the methods and techniques of groundwater investigations. The Clarendon Plains was selected because of the extensive use of groundwater for irrigation, and the abundance of data. Howard Versey of the Geological Survey was selected to receive the training (Versey and Prescott 1958; Van Zandt 1976; J Pennant, *pers. comm., 22 March 2009*). With the support of the USGS, Jamaica was able to generate a large hydrogeological database for the country, along with hydrogeological maps.

In 1963 the Government established a National Water Authority to take over all the large and medium-sized water supplies except those operated by the Water Commission. The Water Resources Section, which at the time was moved to the Water Supplies Branch, formed the nucleus of the authority. Also in 1963, the Government established through the Underground Water Control Act, 1959, a board called the Underground Water Authority. This board was under the Ministry of Agriculture having the Director of Geological Survey as chairman, supported by chief engineers from technical units of the Local Government and the Agriculture ministries, among others (Government of Jamaica 1958b; Government of Jamaica 1965; Jamaica Information Service 1965; Government of Jamaica 1966; Government of Jamaica 1967; Government of Jamaica 1968; J Pennant, *pers. comm., 22 March 2009*; Water Resources Authority 2011).

In 1965, under the United Nations Development Programme's Special Fund, through the Food and Agriculture Organization, a special project (UNDP/FAO project) was implemented to conduct water resources surveys with emphasis on agricultural use for three areas in Jamaica: the Pedro Plains in St. Elizabeth, the Queen of Spain's Valley in Trelawny, and Moneague in St. Ann. The Geological Survey Department was the cooperating partner with the Hydrology and Groundwater Branches assigned. The Water Resources Section of the National Water Authority was transferred to the Survey to work on the project. These entities were also deployed on a hydrographic programme for the Blue Mountain Water Supply Project where gauging stations were established upon request from the National Water Authority and the Water Commission. The National Water Authority had also seconded their Deputy Chief Engineer, Hubert Chin to serve as Senior Hydrologist on the UN/FAO project in 1966. He had worked initially in 1954, at the Water Supplies Branch on the establishment of the first streamflow gauging stations (Jamaica Information Service 1965; Government of Jamaica 1966, 1967, 1968; J Pennant, *pers. comm., 22 March 2009*).

Ultimately, with the completion of the UNDP/FAO project in 1974, a new water resources organization called the Water Resources Division was established. This Division monitored surface and groundwater resources across the island. In 1985, the Water Resources Division was merged into the Underground Water Authority, and by 1995 a Water Resources Act was passed which repealed the Underground Water Control Act and the Water Act. This new act established the Water Resources Authority to be responsible for the management, protection, and controlled allocation and use of Jamaica's surface and underground water resources (Water Resources Authority 2011).

5.10 Earthquake Unit

The 6.8 Richter magnitude earthquake of March 1, 1957, that devastated Western Jamaica prompted the Geological Survey to request seismographs from the Seismic Research Unit at the University College of the West Indies in Trinidad. The new equipment would improve the monitoring of earthquake activity. The request was approved, and by October 1961 a network of seismograph stations was established across Jamaica. Instruments were installed, firstly at the Survey headquarters at Hope, then at the town of Black River, and at the Beverly Land Settlement in St. Ann (Government of Jamaica 1963). The St. Ann station however, did not work as the electricity supply required to maintain the instrument was unreliable. The data collected from the seismic stations were analysed and interpreted back at the Seismic Research Unit headquarters in Trinidad. The unavailability of valuable real-time information was an inconvenience and so it was agreed that an office dedicated to data collection and interpretation should be established at the Science Faculty of the Mona Campus of the University. The instrument installed at Hope, the first modern seismograph in the island, was relocated to the University of the West Indies in 1963 which marked the establishment of the new Earthquake Unit (Government of Jamaica 1965).

5.11 Mines and Geology Division

The Geological Survey Department originally was a temporary activity funded by the CD&W Schemes and was scheduled to be wound up by the late 1950s. However, it became a permanent establishment financed entirely by the Jamaican Government. By March 1960 the functions were expanded over four branches outfitted with technical experts. There was a Geological Mapping Branch responsible for stratigraphic mapping and palaeontological studies. The Groundwater Branch focused on hydrogeological investigations for water supply development. The Geochemical Branch conducted chemical analyses and geochemical prospecting. The Drilling Branch undertook exploratory drilling for investigating stratigraphy, mineral occurrences, and groundwater (Government of Jamaica 1961).

The accommodation at the Institute of Jamaica became insufficient; however a new building designed by Norman and Dawbarn of London was being prepared. Construction of the structure at Hope began in 1960. This would be the headquarters for both the Geological Survey and the Department of Mines, and completion and occupation were scheduled for August 1961. Construction kept schedule and the 35 staff members of the Geological Survey along with those of the Department of Mines moved into the new three-storey building. There was an annex with a library and a Geology Museum which would be open to the public during the weekdays. There was also a storage building, a seismic station, laboratory, dark room, a drawing office and space for the Groundwater and Drilling Branch. Zans was absent from the movement, as in the same month he suffered a heart attack. On September 5, 1961, a second heart attack proved fatal. He never moved into his new office and was not able to see the October commissioning of the seismograph for which he petitioned the Government to acquire (Crooks 1987).

Zans had suggested that there should be displays of the works of prominent contributors to Jamaican geology, such as De la Beche, Barrett, Sawkins, Brown and Matley, and this was the basis for the small Geology Museum. It was also proposed to name this museum after a distinguished British scientist. The Survey staff however, unanimously agreed that this idea should be abandoned, and the museum should be called the 'Zans Museum' in honour of the late director. The museum display was arranged at the front of the library and had showcases of mineral and fossil fuel specimens from all over the world. Most of these specimens were later gifted to the Geology Department at the University of the West Indies where a larger Geology Museum was established in 1965. As the Survey expanded, nearly the entire collection, including the display cases, were donated to the UWI Geology Department (Brown and Langner 2002).

The new offices were officially opened on April 5, 1962, with G. A. Moorehead, the Commissioner of Mines, and due to Zans' death, Chubb acted as Director of Geological Surveys (Government of Jamaica 1963; Crooks 1987). Chubb served as Director until July 1963, and finally resigned from the Service in September. Iorwerth Gwyn Hughes assumed the Directorship until April 1965 when Howard Versey took over as Director. Barry McGrath, Technical Assistant and the first Jamaican employed to the survey also retired that year. By 1969, John Bernard Williams assumed Directorship until 1973 when the department was merged with its neighbour the Department of Mines to become the Mines and Geology Division. This new division was headed by the Commissioner of Mines, Raymond Marcio Wright who was first appointed to the Civil Service as geologist in the Survey in August 1964. (Government of Jamaica 1965, 1967; Jamaica Information Service 1965; Hughes 1973). This merger represented the Government's shift to rationalize the management of the Mining portfolio of the Ministry of Mining and Natural Resources. The Minister, Allan Isaacs, stated that:

> To widen the range of services and to improve overall efficiency, the Geology and Mineral Resources branch of the Geological Survey Department has been combined with the Department of Mines, since April 1st, 1973 to form the Mines and Geology Division of this Ministry. This amalgamation will improve the management of our mineral resources, which form a substantial part of our nation's natural wealth (Hughes 1973).

During that time Jamaica's policies for the Mining Sector were changed and there was greater state participation in bauxite mining. The Government therefore established several state-owned enterprises such as the Jamaica Bauxite Institute, Jamaica Bauxite Mining and the Bauxite and Alumina Trading Company. These served as special purpose vehicles to research bauxite, manage Government shares in the bauxite companies, and trade its off-take of alumina from the local alumina plants, respectively (Government of Jamaica 1965, 1967; Jamaica Information Service 1965; Hughes 1973; Barclay and Girvan 2013).

Before the 1973 merger, the Department of Mines was headed by geochemist, Vincent George Hill who began serving as Commissioner of Mines from 1969. He had first worked with Reynolds Bauxite Jamaica Ltd as a geologist and then chief chemist, and later with the Scientific Research Council (SRC) as geochemist. In the late 1950s while Hill was deployed in Ghana, he was invited by Birtill Lloyd, the Council's Executive Secretary (now called Executive Director), to join the SRC which was being established. This new SRC would conduct scientific research directed towards the exploitation and development of local resources to strengthen the economy (Government Public Relations Office 1960). During that time, the Geological Survey did not have an analytical laboratory, especially to measure water quality, and had sought assistance from the Americans, which during that time was providing technical aid through the USGS in conducting groundwater surveys (Versey and Prescott 1958; Van Zandt 1976). The Geological Survey was desirous that the SRC should be equipped to provide the needed geochemical analytical services, and so a geochemist post was included in the structure of the organization when it started in 1960. The Americans fully equipped the lab to test water quality and Hill was deployed to sample and test the wells across the island. Hill, a clay mineralogist, later procured X-ray diffraction equipment to examine Jamaican clays and conduct comparisons of Jamaican bauxite with those from Guyana, Hungary, and other places. The SRC recruited Carlton E. Davis to conduct the analyses [which he shared in Davis (1973)]. Shortly after, Hill departed to work in the USA. While there, in 1968, the Premier, Norman Manley, acting upon advice of Government bauxite consultant Samuel Moment, invited Hill to serve in Jamaica as Commissioner of Mines. Hill became the first doctoral graduate to serve in this post (V. G. Hill, personal communication 4 June 2009; Hill 1978; Davis 1995). In 1973, he was promoted to Principal Director at the Ministry of Mining and Natural Resources where he supervised the three new mining-related Divisions: The Mines and Geology Division, a Statutory and Regulatory Services Division, and a Mineral Research and Development Division. Hill served in this capacity until 1978 when the institutional arrangement for mining was again reorganized. Hill had obtained geology, chemistry, and mathematics degrees from the University of New Brunswick, a mineralogy degree from Penn State University, and finally a geochemistry doctorate from the University of Toronto in 1956.

Hill may have been one of the most impactful leaders in the Jamaican Mining Sector. While as Commissioner of Mines, he recognized information-sharing gaps and sought to promote the knowledge-sharing, especially for mineral resources. Hill recognized, for example, that technical findings and reports on bauxite were oftentimes kept confidential, whether by companies or the Government. He, along with George W. Brindley, Mineralogy Professor at Pennsylvania University and visiting lecturer at UWI in Jamaica (Newnham 1986), organized the first Bauxite Symposium held in 1971. These bauxite conferences attracted the technical experts who presented on bauxite geology and mineralogy (e.g., Davis 1973 and Hill 1973), processing technology and various advancements made by the companies. The second symposium was convened in 1973 to coincide with 'Natural Resources Week' and was held biennially after that, up to the 1980s. The conference transactions were published through the Geological Society of Jamaica as special issues of their journal (Hill 1971; Rousseau 1973; V. G. Hill, personal communication, 4 June 2009). Hill also had many authoritative publications (e.g. Hill 1983; Hill and Ostojic 1984) which have been further advanced and are still internationally recognized (e.g., Hill and Sehnke 2006). Hill believed that 'knowledge is something to be shared' (V. G. Hill, personal communication, 4 June 2009). He was also instrumental in the establishment of the International Bauxite Association.

The new Ministerial Divisions had specialized functions that were formerly either with the Mines Department or the Geological Survey. The Mines and Geology Division was established first in April 1973, continued administering the Mining Law, the Quarries Law (of 1955) and the Petroleum (Production) Law. The Mining Branch comprised a Mines and Quarries Sections which monitored and regulated the bauxite, gypsum, silica sand and aggregate operations ensuring that mandatory requirements were met. There was also a Safety Section evaluating the working conditions in the mines and processing plants. The entity was also responsible through its Geological Survey Branch, for conducting geological research such as the preparation of geology maps (its Regional Mapping Section), and the assessment of mineral deposits and monitoring of active prospecting licences (Economic Geology Section). An Engineering Geology Section was also established to provide guidance in the form of engineering geological advice to all aspects of development planning (Mines and Geology Division 1975).

The other divisions were established later in November 1973. The Statutory and Regulatory Services Division was established to monitor mining concessions and quarry licences in addition to the monitoring the export of bauxite and alumina. Scientific research on bauxite and alumina along with bauxite reserve quantification was the function of the Mineral Research and Development Division. This Mineral Research and Development Division was reorganized in 1976 to form the Jamaica Bauxite Institute as a subsidiary of the Jamaica National Investment Company (Mines and Geology Division 1975; Davis 1995).

In 1978, the Mines and Geology Division was split into two new divisions: a Mines and Quarries Division, incorporating the functions of the Mines Branch and the Statutory and Regulatory Services Division, and a Geological Survey Division. At that time, the new Petroleum Corporation of Jamaica was established to oversee all oil and gas exploration. Commissioner of Mines, Raymond Wright headed this new entity. Consequentially, Godfrey Perkins became the new Commissioner of Mines. The Geological Survey was headed by a Director of Surveys, Franklyn Jerome McDonald. He however served for two years before leaving to head the new Office of Disaster Preparedness and Emergency Relief Coordination (ODIPERC). Arthur Jelf Selbourne Geddes took over the role.

Under the Government's reorganization and rationalization scheme for the Civil Service the Geological Survey Division was merged in 1996 with the Mines and Quarries Division to form again, a Mines and Geology Division (MGD). Coy Garrett Roache, Commissioner of Mines who had succeeded Godfrey Perkins was retained as head. The new organization, having two branches, therefore had two Deputy Commissioners. Lawrence Henry, formerly the Director of Surveys after Arthur Geddes, became a Deputy Commissioner of Mines having responsibilities for geological matters. Clinton George Thompson became Deputy Commissioner of Mines having responsibility for mining matters. Thompson later succeeded Roache in 2007, when the latter, who had served 20 years as Commissioner, became the chief executive officer of Jamaica Bauxite Mining, the Government's holding company for equity in some of the bauxite companies. The tenures of both men as Commissioners are marked by amendments to modernize the mining and quarrying laws. During the tenure of Roache, the Mining Regulations were amended in 2004 to reflect international best practices in rehabilitation of mined-out lands. During Thompson's tenure the Quarries Control Act of 1983 was amended in 2015 to improve the management and accountability for quarries and their products. Since 2018 the Commissioner of Mines is Roy Fitzgerald Nicholson.

5.12 Petroleum Corporation of Jamaica

Oil and gas exploration in Jamaica in the 1950s had disappointing results. It was thought that with the right expertise a reinterpretation of the data could lead to some discovery success. Therefore, in early 1960s petroleum expert Hugo Greiner was commissioned to reassess all previous works and data. This however did not lead to anything promising and exploration work seemed to wane. The 1973 oil crisis, however, incited renewed interests in Jamaica's potential energy resources and the recognition that the Government needed an entity dedicated to assessing energy resources and guide their development.

In the early 1970s, the Ministry of Mining and Natural Resources had commissioned the exploration of fossil fuel resources such as peat, and reassessment of the oil and gas data. The swamps of Western Jamaica in Negril and Black River were known to host large deposits of peat. Measurements and quality tests were conducted through coring, pilot extraction, and drying experiments in 1978. About 40 tonnes of fuel was shipped to an Irish peat power station which proved the material suitable. However, there were challenges with the drying efficiency and the material was found to burn with similar characteristics to Irish turf. Later surveys and tests showed that nearly 45 million tonnes of peats existed in these swamps and could be used to sufficiently run two small 60-megawatt power plants for 30 years (Robinson 1983; E Robinson, *pers. comm., March 13, 2023*). The Ministry had also engaged expert oil geologist Arthur A. Meyeroff of Tulsa, Oklahoma to assess all the available oil and gas data. He compiled a comprehensive exploration guide, *Petroleum Potential of Jamaica* (Meyeroff and Krieg 1977) in 1977 where he concluded that Jamaica indeed had oil. Meyeroff acknowledged that there were three principal requirements for exploration success. He listed adequate capital, technical expertise, and dedication. He added that Government petroleum organizations were lacking in these compared to the private sector.

This led Raymond Wright to propose the establishment of an independent energy entity to Prime Minister Michael Manley. This entity would collaborate with exploration companies to increase the chances of discovery success (Wright 2010). This resulted in the establishment of the Petroleum Corporation of Jamaica (PCJ) in June 1977; a small organization based in Downtown Kingston. By June 1979, the Petroleum Act established the Corporation in statute to pursue the development of the country's energy resources with exclusive rights to explore for oil and develop these petroleum resources. Wright became the Director of Exploration. The mandate for PCJ was initially for oil and gas exploration along with the procurement of, refining, retailing, and distribution of petroleum products. The PCJ also led in promoting energy conservation, and as an example, in 1983 it constructed in New Kingston, a highly energy efficient high-rise building as its headquarters. Over the first seven years of its existence, measurements show that it used only

50% of the energy of typical office buildings in Jamaica (Robinson et al. 2012; Jamaica Bureau of Standards 1995). This building also utilized marble from Serge Island, in St. Thomas as wall cladding.

In the 1980s, the PCJ, supported by external funding, drilled several test wells onshore (the Hertford, Retrieve and Windsor wells) and deepened one offshore (Arawak). Oil and gas exploration in Jamaica in more recent times focused on finding the offshore reservoirs. The Walton-Morant Licence, an area spanning 22,400 km^2 south of Jamaica, was assigned to companies Tullow Oil, and then United Oil and Gas. This area was found to host potentially 2.4 billion barrels of oil resources (United Oil and Gas 2023). In 2014, Tullow Oil obtained the exploration licence and began their analysis. The company also took United Oil as exploration partner. In 2018, Tullow Oil conducted 3D seismic surveys over 2,250 km^2 in their licence area. In that year, the PCJ announced that two independent live oil seeps had been found onshore in different parts of the island. However, the Government entity was tight-lipped on the locations of these new 'seeps' that suddenly appeared despite years of expert geological work in Jamaica, but oil-leaking farm equipment are a plausible source (Jackson 2019; United Oil and Gas 2023). By the following year, United Oil announced that the possible oil reserves in their licence area are estimated to reached up to 229 million barrels, with a 20% chance of success of striking oil. The company reminded that historically in Jamaica, oil and gas were found in 10 of the 11 wells that have been drilled (Jackson 2019; McDonald 2019). In 2021, United Oil declared that it would cost approximately US $35 million to drill an oil exploration well offshore Jamaica in the 'Colibri' zone where they are most likely to find oil. The Colibri zone is in 750 m of water and has about 406 million barrels. Up to 2021 United Oil had spent US $4.5 million (or J$675 million) on oil exploration in Jamaica. It had also acquired 100% of the prospect from Tullow Oil in 2020, and the Government had extended its licence until the end of January 2024 where they are required to make a 'drill or drop' decision (Jackson 2021, 2022).

In 1995 the mandate of PCJ was expanded to include the development of indigenous renewable energy resources, and it became a main entity in the implementation of Jamaica's National Energy Policy 2009–2030. The corporation held equity in several state-owned companies involved in refining, retailing and distribution, along with renewable energy. Unfortunately for the PCJ, in 2018, the management of the company and some of the subsidiaries were investigated and probed, with adverse findings. Soon after, in April 2020, the corporation was subsumed into its parent, the Ministry of Science, Energy, and Technology, as the Government sought to improve efficiency in management. Of note is PCJ's subsidiary, Wigton Wind Farm which was established as a green energy company in 2000 to provide renewable energy solutions, specializing in wind power. Its success merited placement in 2019 on the Jamaica Stock Exchange. (Linton 2019; MSET 2019; Scott 2019; Saunders 2020; Wigton 2023).

5.13 Office of Disaster Preparedness and Emergency Management

Towards the end of the 1960s Jamaica experienced severe drought. However, in 1979, Jamaica experienced two significant rainfall events, specifically on the western side of the island. In April to May the western Jamaica received rainfall that resulted in significant damages. This was compounded by torrential rains that fell on June 12, 1979, which led to more flood devastation. The Meteorological Division estimated that over 85 billion gallons of water were released on the area over the 24-h period in an event having return period of 50 to 150 years. The event resulted in several hydrogeological and drainage problems. Forty-one persons lost their lives, and there was destruction to houses, bridges, roads, public utilities, and crops. The town of New Market in St. Elizabeth, for several months, was submerged in 80 ft. of water (Patterson 1979).

The relief response by the Government was immediate and support was received both locally and overseas. The bauxite companies actively participated by providing building materials, food, culvert pipes for new drains, and agricultural supplies such as tools, feed, and fertilizers. One company immediately obtained aerial imagery of the disaster area for the Government to use in the response programme. It also allocated J$356,000 (US $200,000) to provide homes for the indigent (Patterson 1979).

The Government established two response groups; an Emergency Operations Centre tasked to undertake and facilitate immediate recovery actions in the disaster area, and a Rehabilitation and Reconstruction Task Force to formulate and implement the reconstruction programme. The latter task force included the Geological Survey Division and the Water Resources Division which were required to conduct assessments to understand the nature of the hydrogeological and drainage problems (Patterson 1979).

It was the intention of the Government that the two response groups should be temporary undertakings. However, the Office of Foreign Disaster Assistance of the United States Agency for International Development (USAID), the United Nations Disaster Relief Office and the League of Red Cross Societies recommended that the Government of Jamaica upgrade the country's emergency management and disaster assistance planning capacity. As a result, the Government in 1980 established an Office of Disaster Preparedness and Emergency Relief Coordination (ODIPERC) as a government entity (Patterson 1979; Conahan 1983).

The ODIPERC, based at the Office of the Prime Minister, was tasked to develop plans to reduce the potential damage from all types of national and localized disasters. This also included building public awareness on disasters to strengthen capacity for survival. Franklyn McDonald, Director of Surveys, was seconded to lead this new office as its first head when it became permanently established (Aratram 1999). ODIPERC was renamed the Office of Disaster Preparedness and Emergency Management (ODPEM) under provisions made in the Disaster Preparedness and Emergency Management Act, 1993. This act was repealed in 2015 and gave way to a Disaster Risk Management Act of 2015 which made detailed provisions for the management and mitigation of disasters, and the reduction of risks associated with disasters (Government of Jamaica 2015; ODPEM 2019).

5.14 Rudist Research

Today, there are about 40 genera and 140 species of rudist bivalves found in Jamaica. These fossil specimens have found usefulness in biostratigraphic correlation (determining and comparing the ages of rocks) and have been applied in resolving the geological history of Jamaica, and as stratigraphic markers in local oil and gas exploration (James-Williamson 2013). In surveying the northern flank of the Blue Mountain in 1860, Government geologist Lucas Barrett found marine fossil bivalves in limestone in the 'Palaeozoic' rocks described by Henry De la Beche in the 1820s. Among the fossils was a well-preserved rudist bivalve mollusc called *Hippurite*. Samuel Woodward of the British Museum described it in 1862 and named it *Barrettia* in honour of the discoverer (Woodward 1862). Barrett's discovery was critical to determining the age of Jamaican rocks where he proved that the older rocks were of Cretaceous age and not Palaeozoic as De La Beche had concluded. Further to this, the overlying red or purple beds of sandstone and conglomerates, and carbonaceous shales could not be upper Palaeozoic, but Cenozoic (De la Beche 1827; Sawkins 1869; Anonymous 1898).

After the death of Barrett in 1862, the palaeontological studies on Jamaican rudists during that Geological Survey continued with George Parkes Wall when he was deployed as geologist to Jamaica during that time. He prepared a collection of 12 species from Clarendon, though these were never listed among the 135 Jamaican fossil species mentioned in the 1869 final geology reports (Sawkins 1869; Chubb 1964, 1980). Other well-known studies on rudists include that by Robert Whitfield in the late 1890s, and Charles Trechmann in the 1920s who also named some (Whitfield 1897a, 1897b; Trechmann 1922, 1924). Amateur scientists also made contributions, such as the discoveries of large rudists during road cutting reported at scientific meetings at the Institute of Jamaica in the 1890s. During the 1920s, Government geologist Charles Matley sent numerous fossils including rudists to the US National Museum in Washington for identification (Anonymous 1923; Stephenson 1942; Chubb 1955b).

In the 1950s, Government geologist Chubb took an interest in resolving the discrepancies of Cretaceous stratigraphy in Jamaica, including correlating the rocks with the rest of the Caribbean. His interest in rudists arose from his analysis of the several hundreds of specimens collected by the Government geologists of the modern Survey during their initial four years of mapping. Specimens described and named included *Praeradiolites verseyi,* after Howard Versey, palaeontologist in the Survey who found the first specimen, and *Sauvagesia mcgrathi* named after the Jamaican technical assistant, Barry McGrath who had also collected the first specimen. Chubb undertook the enormous task of documenting all the known species of rudists in Jamaica. He located Wall's fossil collection in the Natural History Museum, London, and identified a new species (Chubb 1980). In 1954 he devoted a fortnight to study Whitfield's specimens in the American Museum of Natural History, then later visited the National Museum in Washington to examine Matley's specimens. He then proceeded to the British Museum (Natural History) to analyse Trechmann's specimens as well as the specimens of *Barrettia* examined and named by Woodward, and the other Jamaican rudists collected by Lucas Barrett and the cohort of Government geologists between 1859 and 1864. On the museum tours Chubb compared the Jamaican types with those found in Cuba, Haiti, Puerto Rico, the Virgin Islands, Mexico, Texas, and south Europe among other places (Chubb 1955b).

Chubb compiled all his work along with those of the other palaeontologists in a taxonomic monograph *Rudists of Jamaica* (Chubb 1971) where he described and illustrated 71 species. This document was published shortly before he passed on October 12, 1971. In recording the specimens, he affectionately named some after his coworkers and colleagues (Strickland 1989). Chubb's rudist collection was split between the Geological Survey in Jamaica and the Geology Museum at the University of the West Indies (UWI) at Mona, Kingston. His monograph records that these were sent to the Smithsonian Institution in Washington D.C. and the casts of some were placed in the British Museum (Cleevely 1983) However, the type specimens (the first specimen characterized and named) that were at UWI when he documented them were brought back to the Geology Museum at UWI Mona (Brown and Langner 2002).

Along with Chubb's collection, the UWI Geology Museum hosts two other collections of rudists: Lucas Barrett's collection, and that from the Institute of Jamaica. Barrett died on the job in Jamaica in 1862, and his collection of 153 rocks, minerals and fossils was thereafter sent to his family back in England. These were passed to the Sedgwick Museum in Cambridge, and in 1975 acquired by the UWI Geology Museum. The collection of rudists from the Institute of Jamaica were assembled in the late nineteenth century. The labelling was damaged during the 1907 earthquake but was restored in the 1940s (Anonymous 1940, 1941). The collection of British palaeontologist Charles Taylor Trechmann was also donated to the Institute (Anonymous 1948b; James-Williamson 2013). The UWI Geology Museum acquired the collection in 1969; however, these were again to suffer from loss of identity during Hurricane Gilbert in 1988. This resulted in them having diminished value (Brown and Langner 2002).

In June 2011, UWI hosted the Ninth International Congress on Rudist Bivalves in Kingston. This coincided with the 40th anniversary of the publication of Chubb's monograph, and the event was dedicated to his studies of rudists in Jamaica. The UWI continues to advance on Chubb's work and the Geology Museum boasts over 5,000 specimens of rudists, in addition to a collection of over 5,000 images from rudist specimens from the USA, Puerto Rico, London, Cuba, Mexico, and the Netherlands (Mitchell and James-Williamson 2013; James-Williamson 2013).

5.15 Origin of Bauxite

Before 1942, Jamaican bauxite was deemed a material of little value, that is, sub-economic. Over the next ten years and with about J$20 million (US $17 million) expended on mineral exploration and extractive metallurgy, its economic status had been promoted to a metallic mineral of potentially significant financial gain. The occurrence of commercial bauxite deposits in Jamaica as a geological material suggested that similar material could be found in the neighbouring countries. Reynolds Mining Corporation, in 1943, found bauxite in Haiti during their exploration programme. Soon after, in 1944, similar discoveries were made by Alcoa Mining Company in southwestern Dominican Republic. Though smaller that the Jamaican deposits, several companies staked claims (Zans 1952b; Walker 1971; Davis 1989).

The deposits in Hispaniola were examined in detail by the USGS geologists Samuel S. Goldich and Harlan R. Bergquist between 1947 and 1948 (Goldich and Bergquist 1947a; b). In 1942, Bergquist was deployed on a joint programme between the USGS and the US Bureau of Mines to assess bauxite deposits as a strategic mineral. A few years later he was asked to join Goldich of the Foreign Geology Branch of the USGS to examine the deposits in Hispaniola (Brown 1982). They calculated for Haiti, 15 million dried tonnes of bauxite reserves where two-thirds were recoverable. In Dominican Republic the reserve was 6 million tonnes with 2.5 million tonnes being low-grade ore. The Jamaican deposits were documented in successive publications by Reynold's geologist Carl Schmedeman (Schmedeman 1948, 1950), Jamaica Bauxites Limited's geologist, Horace Hose (Hose 1950), and by Zans of the Geological Survey (Zans 1951a). Schmedeman's account was the first to address all the Caribbean deposits. He estimated the total potential reserves of bauxite in Jamaica and Hispaniola to be 350 million tonnes. He calculated that 90% is in Jamaica, with Manchester parish accounting for half of the total reserves, St. Ann parish for one-third, and St. Elizabeth parish and the other parishes in a belt of bauxite deposits for the remainder. By 1950 it was calculated that 130 million dry tonnes of ore were to be found on those properties own or controlled by the three companies in Jamaica. With the vigorous exploration programmes the reserve estimates in these countries increased dramatically. For Jamaica alone, 2 billion tonnes of commercial bauxite were declared (Zans 1952b). The main deposits of course were in Manchester, St. Elizabeth, St. Ann, and Trelawny. The smaller deposits were in St. Catherine, Clarendon, and St. James, and a very minor amount in the John Crow Mountains of Portland parish in the extreme east.

American, E.C. Harder attempted to resolve the origin of bauxite but his research was inconclusive (Harder 1949, 1952). In Jamaica, numerous theories arose to explain bauxite genesis resulting initially, in two theories. The first, called the Residual Theory, was supported by Schmedeman, Hose, Harder, and initially by Zans. They asserted that Jamaican bauxite was the clayey residue of weathering (or the breakdown) of the limestone bedrock and that material was deposited in features such as cavities and pockets within the limestone developed through karstification. That material was subsequently altered. I. G. L. Sinclair in 1967 calculated that 800 ft. of limestone would have had to be dissolved to give the quantity of bauxite and terra rossa deposits seen in Jamaica (Sinclair 1967). The other theory, later held by Zans as an Alluvial Theory, proposed that river-deposited material, originating from the older rocks in the central parts of the Island, was transported into the limestone areas and subsequently altered to bauxite by weathering. The idea was based on the sharp limestone-bauxite contact observed in the deposits. This theory however, assumed that Jamaica looked the same as it is today, when the bauxite deposits were being developed

(Zans 1959). However, Goldich and Bergquist from the late 1940s had proposed that material ejected from a volcano was the original material that had broken down to form the bauxite. Their work, however, was done in Hispaniola (Comer 1974).

John B. Comer of the University of Texas at Austin resurrected the idea of Goldich and Bergquist in what became known as the Volcanic Ash Theory (Comer 1974). Using mineralogy and chemistry he showed that the host limestone did not have sufficient alumina to reduce to bauxite, which refuted the Residual Theory. He further added that the thickness of limestone weathered to produce bauxite was incorrect as it did not consider that erosion at the time would also remove some bauxite. He therefore supported that volcanic ash deposited on Jamaica and Hispaniola from an ash cloud during the Miocene times served as the source of the bauxite deposits. He also mentioned that there is a corresponding layer of bentonite clay in the Jamaican Miocene white limestone formed from some of the volcanic ash deposited in the ocean during that period of eruption. The Volcanic Ash Theory was accepted as the most plausible. The source of the ash was not defined until in the 1980s when satellite imagery showed an ash cloud originating from a volcanic eruption in Central America drifting towards the Leeward Islands of the Caribbean (Lyew-Ayee 1986).

5.16 Geological Mapping

Since the 1860s the geological maps produced by the Jamaican Geological Survey were considered the basis to support mineral exploration, water resource assessment, agriculture, and engineering projects. The major mapping programme of the 1950s benefitted from the new rock exposures and the abundant geological data made available during large-scale physical development activities such as port development, tunnelling projects, and road cutting. The mining of bauxite and the drilling for oil also provided access to subsurface information.

Map production by the survey was given a boost in 1968 through technical cooperation with the British Geological Survey (BGS). A rigorous programme was conducted to map, compile, update, and symbolize the geology of Jamaica at the 1:50,000 scale across 30 sheets based on the geological map of Jamaica in 1958 (Zans 1958). This long-term cooperation, the Overseas Technical Assistance Programme, was funded by the British Ministry of Overseas Development. BGS Staff were deployed in Jamaica for several years undertaking major field work with local ‘counterpart’ geologists. Apart from Jamaica, countries such as Iran, Bolivia, Indonesia, and Ecuador benefitted from similar assignments lasting from six to eight years, most of which started in the mid-1970s. At its peak the programme was being implemented in 30 developing nations (Wilson 1985).

In Jamaica, the cooperation lasted until 1973, with experts such as stratigrapher Gilbert W. Green to oversee the mapping. Other geologists included Iorwerth Gwyn Hughes and J. H. Bateson. Hughes, a former Director of Geological Surveys in Jamaica in the early 1960s, was tasked to compile information on mineral resources (Hughes 1973), while Bateson compiled the first 13 or 14 maps which were printed in black and white between 1972 and 1974. During the programme, Jamaican, Anthony R. D. Porter joined the survey in 1971 to map the Kingston district. When the BGS cooperation ended, Porter was tasked to produce the final maps. Three maps of the series were coloured to represent the greater accuracy of data gathering compared to those in black and white which utilized aerial photography in most places, substantiated by academic field surveys (Government of Jamaica 1995). By 1974, Porter joined Alcan Jamaica Limited but with special permission from that company was able to continue finalizing the maps on which he was deployed. Alcan also permitted the inclusion of the findings from their geological projects in the national maps. This included the results from their geological and geotechnical assessments conducted in early 1972 when the company considered using a direct cable belt conveyor installed overland and underground and passing through three tunnels to transport bauxite from its future reserves, located in Pedro Valley southwest of Claremont, St Ann approximately 19 km (12 miles), to the Ewarton alumina plant. In early 1976 Alcan abandoned the project on the basis that it was uneconomical when the cost was considered and coupled with the burden of the newly introduced bauxite production levy (Porter 2020).

Along with the geologists of the BGS, scholars and students of the Geology Department of the University of the West Indies contributed information for inclusion on the national geological maps (Bhalai 2012). The island was in most places mapped geologically but the more detailed medium-scale (1:50,000 scale) sheets were not consistently released. During the time of the British technical cooperation, 20 sheets (of a total of 30 sheets) were produced (British Geological Survey 1991). These covered two-thirds of the island. Up to 1978, five additional sheets were prepared. Preparation of those remaining five sheets which represented the eastern side of the island from Swift River in Portland parish, extending around the east coast to Yallahs in St. Thomas, was sluggish though it was reported that these maps were compiled in the late 1970s but were never published, possibly due to incompleteness (Government of Jamaica 1995). The mapping programme benefited from the service of volunteers from the US Peace Corps who helped to expedite the preparation. In the early 1980s there were 45 persons on staff at the Geological Survey Division which included mainly Caribbean nationals and periodic attachments from international technical schemes. The geological maps finally became available for printing in 1987; however some lacked key elements that added value, such as topographic contours and explanatory margin notes of the geological formations (e.g., GSD 1987a, b). In addition to the medium-scale maps, a 1977 island-wide compilation of the geological mapping titled *Jamaica Geology* (1:250,000) was also prepared (McFarlane 1977) and was later revised in 1984 (Fig. 5.5; Brookes 1984).

Reproduction of the geology map series encountered delays from their scheduled 1987 release date due to unreliability of the printing service at the Government's topographic survey office which was commissioned to make the copies. Fortunately, with overseas assistance, all the maps except for the eastern sheets were printed in quantities and became widely available in 1989 (Government of Jamaica 1995).

5.17 Special Research Programmes

National geoscience entities have the core mandate to conduct practical research to meet the current and projected development needs of their country. In Jamaica, several notable market-driven research programmes have been implemented. Occasionally, like the mapping support given by the BGS in the 1970s, international technical assistance and funding have contributed immensely to advance this applied research.

5.17.1 Engineering Geology

In 1979 a formal study of the general engineering properties of Jamaican rock types was initiated to understand bedrock behaviour to inform and guide physical planning and development. It was intended that local development projects would be equipped to make preliminary decisions towards addressing potential engineering problems. Rock formations were therefore grouped into according to geotechnical behaviour, and these were presented in a 1983 compilation, *A Geotechnical Classification of Jamaican Rocks* (O'Hara and Bryce 1983). This document was complemented with an island-wide (1:250,000) geotechnical map of Jamaica (GSD 1984).

Slope stability analyses became a core activity of the Survey primarily due to Jamaica having large expanses of hilly country. Support was provided to the Forestry Industries Development Company (FIDCO) in 1985 to guide forest

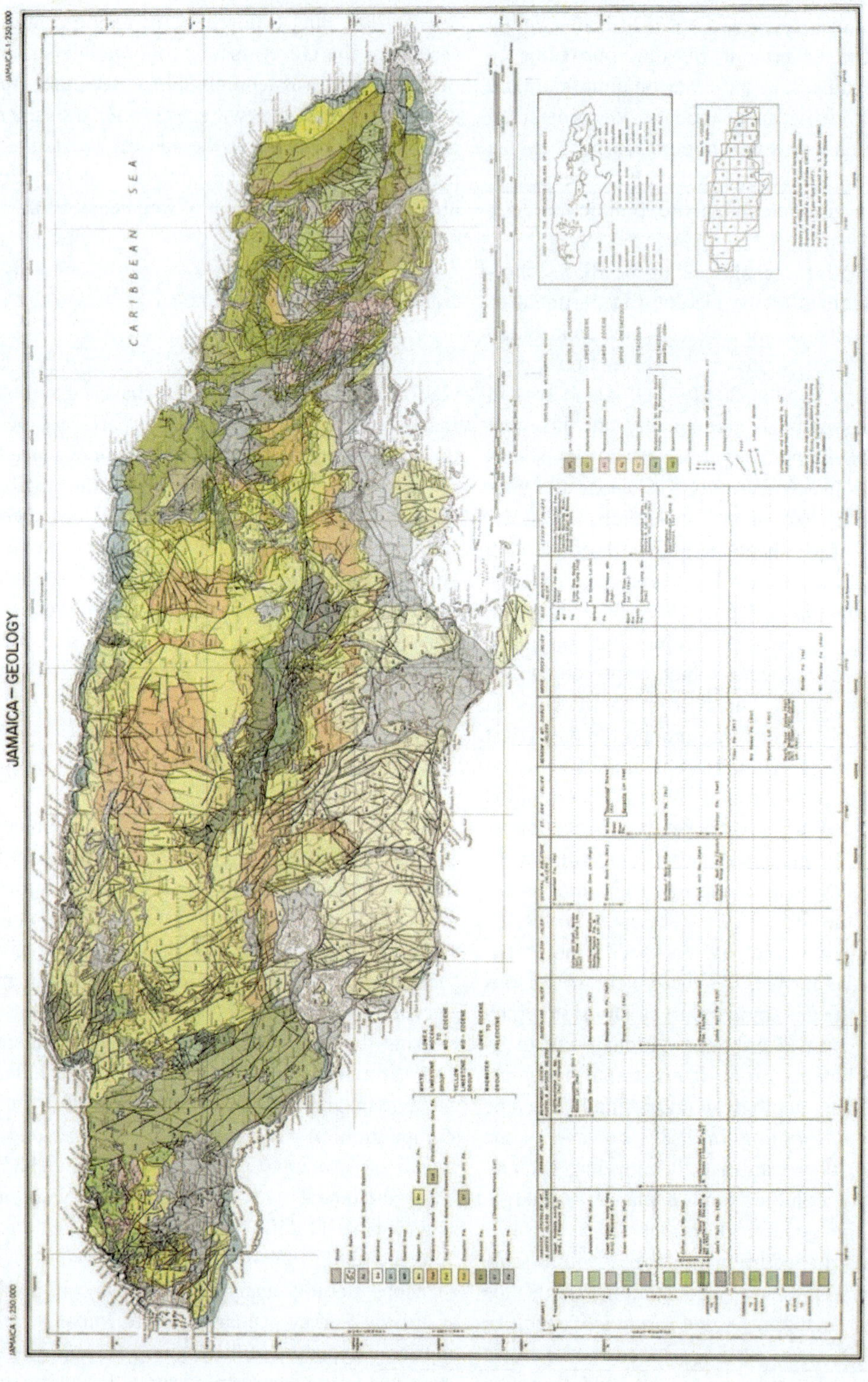

Fig. 5.5 1984 Geological Map of Jamaica (Brookes 1984) updated from the 1977 edition (McFarlane 1977) (Courtesy of the Mines and Geology Division, Jamaica)

road cutting on the southeast flank of the Blue Mountains. This Government company had the primary role to produce forest products from plantation-grown pine logs. By determining specific geological conditions in the sensitive mountainous terrain, stable areas on the slopes could be identified for forest road construction and to reduce environmental degradation from landslides and erosion. Early in 1986, the Survey was deployed at Preston, St. Mary, to investigate a large slow-moving landslide. The movement was continuous and spanned a large area where buildings were being distorted and destroyed as the ground slowly shifted. By May 1986, upon recommendation of the Survey, 17 families were relocated. The assessment continued into 1987 when boreholes were placed at several points within the landslide to understand the movement. The testing revealed that limestone bedrock was sliding along the underlying clay formation, a theory that was proposed by English geographer Vaughan Cornish in 1907 when the Kingston Earthquake had occurred (Anonymous 1907; Bryce et al. 1987).

In the late 1990s the Survey revived its programme of examining landslides by developing predicative models of landslide susceptibility. The heavy rains in January 1998 had triggered many disastrous landslides that resulted in over US $6.3 million (J$230 million) in damages in housing, infrastructure, agriculture, and other sectors, in addition to causing five deaths (Harris 2002). This prompted a pilot-scale landslide survey in that year where a prediction model was developed for the Rio Grande Valley in Portland parish. This valley between the Blue and John Crow Mountains has a history and reputation for damaging landslides and it was therefore fitting to conduct this assessment. It was intended that the susceptibility model would build resilience of the communities and reduce the impact on lives and property. The landslide survey was funded by the European Commission Humanitarian Aid Department's Disaster Preparedness Programme (DIPECHO). The landslide susceptibility model prepared from this assessment was the first computer-generated susceptibility model prepared by the Survey and utilized analytical techniques perfected by Italian geomorphologists.

The landslide susceptibility assessment programme continued with surveys of St. Thomas parish in 2001, supported by Government funding. St. Mary parish, well-known for landslides, was surveyed in 2002, through grant funding from the United Nations Development Programme (Bhalai and Harris 2004), and a complete assessment of Portland parish through Government funding in 2006 (Bhalai 2007; Bhalai et al. 2007). The map of St. Thomas was refined in 2009 (Bhalai 2009), and an assessment of St. Catherine parish was also undertaken some years later. The susceptibility analysis technique employed by the Survey was adapted from those used in North America and Europe, and customized to model the Jamaican environment (Bhalai 2010).

5.17.2 Mineral Exploration and Development

Since its establishment, the Survey has served as the country's repository for information on all the known mineral resources. This knowledge base was publicized with the release of a comprehensive compilation on the *Economic Geology and Mineral Resources of Jamaica* (Zans 1951a). During the 1960s up to the time of the BGS technical cooperation, mining in Jamaica was at a peak, with Jamaica being a top performer in global bauxite production. In addition, silica sand was being produced from the Black River basin for the West Indies Glass Company. Dimension stone, especially marble was being exploited in the Serge Island area of St. Thomas, and a small stone craft industry was developed at Hellshire Bay (United Nations 1988). Iron ores in St Andrew parish were also being explored. The abundance of new information merited an update which was undertaken by Iorwerth Gwyn Hughes in 1973. This resulted in *The Mineral Resources of Jamaica* (Hughes 1973) being published. Additional information was collected on metallic minerals such as copper and gold deposits, and energy resources such as peat in Black River and Negril, and the 1973 bulletin was revised and updated in 1981 (Fenton 1981). In addition to this promotional volume, a bulletin showcasing 31 *Mineral Springs of Jamaica* was published in 1987 (Hylton et al. 1987).

When the Geological Survey Division was established in 1978, the Government had the development objective to accelerate the evaluation and rational expansion of the country's non-metallic minerals subsector. A unit was therefore established between 1978 and 1979 dedicated to examining these industrial minerals. That unit initially focused primarily on the development of limestone deposits to supply the demands for road and building stone at the major town centres across Jamaica. At Montego Bay for example, a 'quarry zone' was demarcated between 1983 and 1984 where suitable stone could be found to support development of that second city (Tyson 1983). Minerals such as marble, silica sand, and high-purity limestone unfortunately were not prioritized, probably due to a lack of technical expertise in evaluating these high-value commodities.

By the early 1980s, the Government, through the Survey, in conjunction with the United Nations Development Programme's (UNDP) Department of Technical Cooperation for Development conceived a programme for exploring and developing industrial mineral resources. This project commenced in July 1984 and was initially extended to December 1986. The total spending during this 30-month period just exceeded US $1.2 million, with 40% of the funding coming from the Government, the rest largely from UNDP, and a marginal amount from the United States Agency for International Development (USAID). The programme was extended into 1988 throughout the duration, deposits of high-purity limestone and dolomite, and attractive marbles were discovered. Silica sand, gypsum, clay, and stone for construction were also evaluated. The Survey also received physical support in the form of a chemical laboratory, as at that time there was no dedicated analytical facility for minerals. In addition, the organization in 1985 received its first computer, an IBM desktop which supported the establishment of a 'geoscience and mineral resource data storage and retrieval system' for the published Government technical reports. Twelve staff members from both the Survey and the Mines and Quarries Division were trained in its use. The programme also recruited various technical specialists such as electronic data processing, mineral marketing and structural geology experts from the USA, and a marble expert from Carrara, Italy. To strengthen the programme and accelerate the promising progress made, the Government made a request at the Swedish Agency for International Technical Cooperation (BITS) for Swedish experts to assist in further evaluation of the non-metallic mineral resources identified. Between November 1987 and May 1988, the Swedish Geological Company was deployed in Jamaica. This organization specifically examined deposits of high-purity limestone, commented on the extraction of rare earth elements from red mud residue derived from bauxite refining, and provided insight to develop the other minerals highlighted during the earlier phases of the programme (Government of Jamaica 1987; Gustaffson 1988; United Nations 1988).

The UNDP-funded programme successfully resulted in several extractive operations being started. Eight high-purity limestone deposits including those at Epworth and Orange Park in St. Ann parish were identified in economic zones spanning 10 miles from the coast; a distance deemed profitable if the material is to be transported. Four of these deposits were staked and explored further by foreign investors. Seven marble deposits were also studied across the island. The marble tile and craft facility at Hellshire, St. Catherine was expanded to cut and polish both local as well as imported stone from Cuba. This, coupled with a pilot dimension stone plant erected at the Survey, resulted in a small industry being established. This facility had stone block extraction and transporting equipment, large slabbing and trimming saws, and polishers. Marbles were sourced from various deposits across the island and the products were sold on the Jamaican market to outfit many new buildings with decorative stone. It offered prospective investors the opportunity to view the variety of Jamaican commercial marbles, their physical properties and the best practices for extraction, processing, and property-enhancing treatments. Jamaican stone was tested and exhibited in Italy, Cuba, Yugoslavia, USA, and Canada, with high acceptance on these markets (Gustaffson 1988). A catalogue of *Jamaican Marble* (Richards et al. 1998) showcasing 21 local stones was prepared in 1998. This was later updated and revised in 2011 to include maps and additional datasets to improve 'investor friendliness' (Bhalai 2011).

In 2001 the Government of the Czech Republic issued a decree for the *Development and Industrial Exploitation of Non-metallic Resources in Jamaica* to be implemented for ten years. Over €5 million was approved for this 'Czech project' where many non-metallic mineral deposits with the potential for rapid economic development were evaluated using modern geological methods. By 2009, several deposits of high-purity limestone, volcanic rock for skid-resistant aggregates, and clay deposits for use as cement raw material were evaluated. The accompanying knowledge and technology transfers were also beneficial to the Survey staff. Several persons received training in modern field evaluation techniques for industrial mineral deposits, and there were also study tours of mineral operations in the Czech Republic.

The skid-resistant aggregate deposits evaluated under this Czech project represent six bodies of volcanic rocks identified in the late 1990s which warranted further attention. Responding to the demands at that time the Survey began identification and assessment of these rock bodies. Around the mid-1990s contractors were invited to bid for the major rehabilitation work of the Norman Manley International Airport in Kingston. This included resurfacing of the runway and the parking apron. In December 1996 it was announced that a Spanish company had the successful bid. They would implement the project over a period of over 13 months in a contract valued at US $18 million. The company, however, disclosed that road stone from local sources would not be utilized in the surfacing, but 156,000 tonnes of crushed granite from Ramsen Aggregates based in Nova Scotia, Canada, would be purchased. Adding to this, the material would cost 80% to 90% more than local crushed stone (Anonymous 1997a).

Traditionally, Jamaican river shingle was crushed and used in road surface pavements. In the 1700s hard beach shingle from the Palisadoes tombolo of Kingston was used for road works. This was material washed down from the Hope River and transported to the beach by the sea (Steers 1940). The Yallahs River in St Thomas parish offered a similar and more sustainable supply, and by the twentieth century this river stone from was preferred for surface pavements on many major roads. It was also used in the surfacing of the runway of the Norman Manley airport when it was constructed in the 1950s (Anonymous 1997a). Professional groups of engineers and builders were displeased with the disclosure and were unconvinced that Jamaican stone was unsatisfactory. They further stated that the proposal was another attempt to deny local contractors access to large projects. The company spokesman reassured that local material which accounts for half of the amount of aggregate needed for the project would be used in the underlying pavement structure. However, based on the contract specifications, the Canadian stone was the better option for the surface as material failure during usage of the airport could be detrimental. He further clarified that although the stone will be acquired at a higher cost, the company will absorb this expense. The aggregate would cost approximately CAD$4 million (Douglas 1996; Anonymous 1997a, 1998).

The disapproval of the engineers and builders group prompted the transport minister Robert Pickersgill to issue a statement that he was certain that no local contractor was able to process the sand and stone needed 'to meet the quality, quantity and … the required time frame'. He however explained that the importation would not be sufficient to meet the aggregate needs and an additional 200,000 tonnes of material valued at J$50 million (US $1.4 million) would be obtained from local suppliers. The construction groups challenged his rebutted statement citing that Jamaican aggregate was, at that time, being used to refurbish the US-owned Guantanamo Bay airbase in Cuba and was therefore capable of being used at the local airport. The minister however clarified that the Jamaican material used in Cuba was being utilized in the pavement structure and that the surfacing aggregate was from the same Canadian source (Anonymous 1997b).

In March 1997, a senior civil and structural engineer who had worked extensively in North America refuted the statements of the Minister and the contractor. He explained that one of the companies that worked at Guantanamo Bay in 1993 had reported using between 60,000 and 70,000 short tonnes of aggregates in the asphalt pavement course including the surface pavement, and these were supplied by Caribbean Aggregate Jamaica and Jamaica Premix. These were two sand and gravel operations established in 1992 and 1994 to crush and screen river shingle sourced from the Yallahs River. This company also stated that the US airbase had the most stringent specifications and utilized standards from the American Society for Testing and Materials (ASTM). Caribbean Aggregates, in January 1997, had earlier supplied aggregate for an airport construction project in Nevis. The first shipment of 52,000 tonnes of Canadian stone arrived at Port Royal in January 1997 with similar sized loads over the next two months. The runway was rehabilitated to accommodate 10,000 movements per year for the next 20 years. At the inception of the project the movement activity was 1,500 movements per year (Anonymous 1997a, b; Martin 1997).

In late 1998 Pickersgill had become the new Mining Minister, and he announced that tests were being conducted by the newly established Mines and Geology Division, on hard igneous rock bodies in St. Thomas and other places for skid-resistant aggregates (Anonymous 1998). Sometime later it was reported that there were six deposits identified having millions of tonnes of stone having similar or better properties than that from Canada (Anonymous 1999). Unfortunately, the deposits were never developed to production until the mid-2010s when one hard rock body in northern St. Catherine parish was used by Chinese contractors in the North–South Highway project. The contractors started using the stone for the pavement structure and were trucking Yallahs River shingle for the surface pavement. During a tour of the project the MGD suggested the use of the hard rock for the surfacing to substitute the river stone. Despite this one-off and successful use, rock bodies for skid-resistant aggregates remained largely underused and are classified as an economically important 'Development Mineral' by the UNDP (Franks et al. 2016).

This recognition that several industrial mineral deposits including those for skid resistant aggregates remained underdeveloped led to the ACP-EU Development Minerals Programme considering Jamaica for technical assistance. This capacity-building programme started in 2016 and focused on building the importance and improving the management of industrial minerals, construction materials, dimension stones, and semi-precious stones. The programme, an initiative of African, Caribbean Pacific (ACP) Group of States, was coordinated by the ACP Secretariat, financed by the European Commission and UNDP, and implemented by UNDP. Jamaica has since benefitted through the training of hundreds of individuals in Government and the private sector in many different practical aspects in mining. The MGD received support towards strengthening its spatial data infrastructure to make its information service digital and customer centric (Bhalai 2018). This included the development of a digital mining cadastre for real-time information on exploration and mining tenements such as those for high-purity limestone, bauxite, copper, and gold.

The modern Survey, since its inception in the 1950s, undertook geochemical surveys and prospecting work to identify metallic mineral resources. The findings greatly assisted the many exploration companies who showed interest in the metalliferous districts that have primarily copper, gold, and iron resources. Eastern Jamaica was considered highly metalliferous and was the focus of the United Nations Blue Mountain Drilling Project of the 1970s. Places with abundant showings such as Ginger Ridge and Castleton in St. Andrew parish were drilled and assayed, and some samples yielded up to 20% copper (Black et al. 1972). Historically, non-bauxite metallic mining in Jamaica was discouraging, and not surprisingly, the findings from the drilling programme were never advanced. In the late 1970s, Vincent Hill, former Commissioner of Mines, prompted the collation of data from all private and Government-funded copper exploration. He recognized that the previous copper exploration work was generally considered disorganized, and the compilation and review would be of great assistance to the prospector. This assignment also benefitted from the assistance of the French geological survey, the Bureau de Recherches Géologiques et Miniéres (BRGM) which deployed geologists to assist with field studies and provided economic interpretations. This resulted in the *Copper Prospects of Jamaica* becoming available as a comprehensive guide (Fenton 1979).

In 1985 the Government of Canada through its Canadian International Development Agency (CIDA) entered a technical cooperation agreement with the Jamaican Government, represented by the Geological Survey, to conduct studies to support metallic mineral exploration and development. The 'CIDA project', as it was branded, was valued at CAD$5 million, and largely involved island-wide geochemical and geophysical surveys with some geological mapping and test drilling. These surveys investigated the potential for non-ferrous base and precious metals across Jamaica. A geochemical database was prepared and areas with potential for mining were also identified. The database would also serve to guide national development of the Mineral Sector and provide other sectors such as Health and Agriculture with an understanding of the geochemical conditions in the natural environment. The project resulted in a transfer of modern metallic mineral exploration techniques to the Survey (Garrett 1985; CIDA 1988; Government of Jamaica 1995).

The CIDA project was divided into three implementation phases. The first phase spanned the first ten months of 1986 with an island-wide geochemical survey of nearly 2,000 km^2, and the collection of over 3,000 samples of rock and soil. These were analysed at a laboratory in Ottawa as the Survey did not have adequate laboratory facilities. This assessment resulted in 210 areas of importance being identified with 17 having high potential for gold. The second phase started in late 1989 with specific geochemical work along with geophysical and geological surveys in the high-priority areas. However, on this occasion the analyses were done in Jamaica at a fully equipped laboratory opened at the Survey in March 1991 at a cost of J$2.49 million (US $206,000). Analytical work was shared with the Nuclear Research Centre, another laboratory specializing in geochemical analyses, based at the University of the West Indies where CIDA had provided CAD$70,000 worth of equipment. The findings from the surveys were particularly encouraging and merited that a small delegation attend the popular annual Prospectors and Developers Association Convention at the Royal York Hotel in Toronto, Canada in April to May 1992 to promote the findings (Government of Jamaica 1995).

In the third phase, exploratory drilling of the potential deposits identified during in the second phase along with the quantification of respective ore reserves was planned. However, before this phase could be implemented the project came to a premature end in 1994 due to disharmony between the local and the foreign teams. Along with the Survey, the Public Works Department and the Forestry Department had concurrent CIDA-funded projects. Unfortunately, the project with the Public Works Department also ended prematurely for similar reasons. The working relations at the Forestry Department were more harmonious resulting in full successful implementation. Forest reserves were surveyed by remote sensing methods which included the capturing of coloured aerial photographic coverage of the entire island between 1991 and 1992. Interestingly, this aerial photographic coverage has been an invaluable dataset that supports the Survey's landslide susceptibility modelling projects.

During the CIDA-funded geochemical survey high readings for gold and silver were recorded at Main Ridge, a few kilometres from the village of Pennants in Clarendon parish. In 1989, Jamaica Mining Company Limited initially conducted prospecting but passed on the property to BHP Minerals International Exploration, Inc. This company conducted further soil geochemical tests and surface rock sampling along with a drilling programme of 38 exploratory holes. BHP concluded that the reserves were too small to merit further work, and by early 1993 it was transferred to Orvana Minerals Corporation which acquired full interest in the prospect. Up to 1996 Orvana conducted exploration which revealed 95,725 tonnes of gold ore having an average grade of over 9 grams per tonne (g/t). This was calculated to about 27,739 troy ounces of gold at an initial workable mining limit (Business Information Group 1993; Siriunas 1993; Williams 2006; Exploration Management Services Pty Ltd 2014). Orvana, through its subsidiary, Clarendon Mining Limited, in 1996 entered a joint interest arrangement with Ausjam Pty Ltd, a private Australian junior mining company. However, in June 2000, Orvana sold its interest to Ausjam and its Jamaican subsidiary, Ausjam Mining Pty Ltd for US \$23,000 plus a 2% net smelter return royalty up to US \$60,000 (This royalty paid to a mine-property owner is a specified percentage of the gross revenue from a mine after deducting the incidental costs such as transportation and refining) (Orvana Minerals Corporation 2005; Royal Gold Inc 2014; Williams 2006). Ausjam obtained full rights to the Pennants deposit. The company projected that there were approximately 31,000 troy ounces with substantial silver.

Construction of the Pennants Mine and crude smelting facility began in 1999, and by March 2001 production had started. Ausjam had invested US \$7.5 million. The mine lasted for little over three years before closing in 2004. It had produced 16,540 troy ounces of gold and 9,520 troy ounces of silver (Fig. 5.6). Apart from employment and the resulting commercial activities corollary to the mine, Jamaica gained a mere J\$7 million (US \$122,000) which represented the 4.25% royalty assessed (Williams 2006). The Pennants Mine is the only high-priority area identified during the CIDA geochemical surveys that was ever successfully developed. It represented Jamaica's first modern gold mine.

Since the inception of Jamaica's bauxite industry over 150 million tonnes of red mud waste have been generated (Bhalai and McKoy 2020). It was recognized that the waste does not recycle into the environment but remains in the respective holding area. In the mid-1980s, researchers at the University of the West Indies sought to characterize the red mud to find some useful application of the material, or to identify any potentially valuable constituent (Wagh and Pinnock 1987). The researchers identified that Jamaican red mud had significantly high concentration of rare earth elements. This collection of 17 elements includes the 15 lanthanide elements, yttrium and scandium, and finds application in advanced technology such as electronics, communications, green energy, and defence. The Bayer process concentrated the rare earths originally found in the bauxite, by two and a half times in the red mud (Porter 2017). The bauxite companies were also aware of the potential and, in the case of Alcan, managed to secure a patent in 1991 for an acid digestion process to extract the rare earths from the mud (Fulford et al. 1991). The JBI, between 1991 and 1993, did basic leaching tests and solvent extraction studies but was hindered by a lack of sensitive laboratory equipment to measure the rare earth element concentrations (Barnett 2020).

Since the mid-2000s, the annual demand for rare earths began increasing exponentially. China accounts for 50% of the world's reserves, and around 2010, that country provided about 98% of the world's supply and controlled the sector completely. As early as 1990, the Chinese

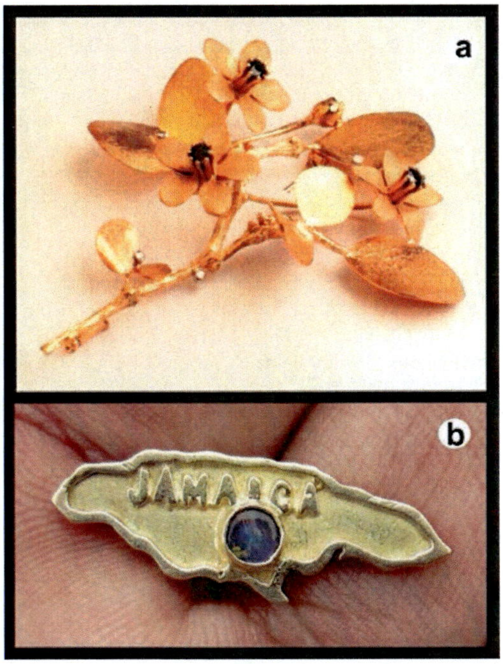

Fig. 5.6 Jewellery made from Jamaican gold mined at Pennants, Clarendon. **a** – 'Queen's Mother's Brooch' presented as Jamaica's gift to Her Majesty Queen Elizabeth, the Queen Mother, for her 100th birthday on August 4, 2000. The brooch is approximately 5 cm × 5 cm and contains 15 g of 18 karat gold. The design was inspired by Jamaica's national flower, the Lignum Vitae. A sapphire was placed in each flower to highlight the blossoms' blue colour. Four diamond dewdrops top off the design which was presented in a box made of wood from Jamaica's national tree, the Blue Mahoe. The brooch was prepared by a team of expert jewellers and craftsmen comprising Carol Campbell, Peter Bangerter, and staff of Swiss Stores, Kingston. The brooch belongs to the British Royal Family and is kept in the Treasury Collection at Buckingham Palace (Courtesy of Jamaica Promotions Corporation [JAMPRO], Kingston). **b** - Lapel pin (2.5 cm long) made of 14 Karat gold shaped like the island of Jamaica, with Australian blue opal marking the site of the Pennants gold mine operated by Australian company, Ausjam Pty Ltd. The pin was prepared by Peter Bangerter (Photograph by author)

Government had declared rare earths to be protected and ranked as a strategic mineral, like gold, and foreign investors could not operate mines or process rare earth ore unless in a joint venture with Chinese firms. Even domestic companies had to obtain state approval including that from their Rare Earth Office. Social media use increased sharply in the 2000s, and this drove the demand for electronics in both the developed and developing countries. This further drove the demand for rare earths and the world began to develop other sources. In 2009, China imposed a six-year export quota on rare earth minerals limiting annual export to less than 35,000 tonnes. There was also a strict prohibition on the export of four of these minerals. This caused a hike in their prices. China had also bought 10 million tonnes of rare earths to add to their stockpile which kept the prices high. Throughout the world more research began into rare earth sources other than China. The ban also triggered the USA to refocus their attention on what minerals are deemed 'critical' (Tien 2013; Anonymous 2014a; Gleason 2014; Adetunji 2021).

Several Asian companies had developed extractive technology, including Tokyo-based Nippon Light Metal Company which had also developed various acid-leaching methods and later sought and obtained patents (Sugita et al. 2014, 2016). In 2010 Nippon started examining Jamaican red mud for its potentially economic REE content (Barnett 2020). In 2012 the company approached the Jamaican Government with a research proposal to conduct detailed tests on Jamaican red mud using their proprietary extractive technique. By September 2012, Nippon entered a Memorandum of Understanding with the JBI to construct a US $3 million REE pilot plant facility at JBI's Hope headquarters to ascertain the commercial viability of extracting the elements. Nippon would ultimately establish a facility to extract approximately 1,500 tonnes of REE oxides per annum (Paulwell 2013). Construction began in February 2013 and the plant was officially opened on October 23, 2013, by the Mining Minister Phillip Paulwell. It was intended that over the five months pilot phase, 30,000 tonnes of red mud would be tested and the results from the trials would be owned jointly by Nippon and the Jamaican Government. If the trials were successful, the Government would begin commercial negotiations with Nippon (Anonymous 2013; Megginson 2013).

By August 2014, it was announced that Nippon's REE project was on hold as it was not economically viable. The World Trade Organization had ruled that China's export quota on the minerals was illegal resulting in a resurging of Chinese supplies on the world market and a corresponding fall in prices (Anonymous 2014a). Nippon later abandoned the project and handed over the plant to the JBI in October 2014. The company had expended J$600 million (US $5.3 million) (Anonymous 2014b). Seven years later, it was announced that several North American companies were partnering to extract REEs and other valuable commodities from the red mud disposal site for the Gramercy alumina plant in Louisiana. The companies, one of which was the parent entity for the private company in the Discovery Bay, St. Ann bauxite mining venture, would invest US $800 million to extract REE from the 35 million dry-tonne reserve of red mud at Gramercy. A significant amount, if not most of this red mud was produced from bauxite originating from Jamaica (Anonymous 2021). This prompted the Jamaican Government to begin exploring its legal rights to the metals found in the red mud at Gramercy (Spence 2021). Since the early years of the bauxite industry, the Government had adopted the policy where Jamaica would be transformed from a mere supplier of raw materials into a provider of finished products. This meant that, as far as possible, raw materials should be improved in Jamaica by the companies, before export (Lightbourne 1970; Walker 1971).

References

Adetunji J (2021) Demand for rare-earth metals is skyrocketing, so we're creating a safer, cleaner way to recover them from old phones and laptops. The Conversation, 16 April 2021. https://theconversation.com/demand-for-rare-earth-metals-is-skyrocketing-so-were-creating-a-safer-cleaner-way-to-recover-them-from-old-phones-and-laptops-141360. Accessed 24 Mar 2023

Anonymous (1898) Institute of Jamaica: Museum Note—Jamaica palaeontology. Daily Gleaner, 11 Mar 1898, np

Anonymous (1907) The Kingston earthquake in Jamaica: its causes and effects. Daily Gleaner 21 Jun 1907, p 9

Anonymous (1923) The report of Dr Matley on his work on the progress of geological surveys in Jamaica for economic purposes. Daily Gleaner, 5 Jul 1923, p 13

Anonymous (1924) Dr Matley's report on our water resources. Daily Gleaner, 30 May 1924, p 6

Anonymous (1940) Extracts from Annual Report of Institute. Daily Gleaner, 29 Jan 1940, p 16

Anonymous (1941) Institute report records progress over long period. Daily Gleaner, 11 Oct 1941, p 12

Anonymous (1943a) Few minutes session of Council. Daily Gleaner, 31 Mar 1943, p 6

Anonymous (1943b) Jamaica prospects. Daily Gleaner, 29 Sept 1943, p 6

Anonymous (1944) Report of the Standing Finance Committee. Daily Gleaner, 17 May 1944, p 10

Anonymous (1945) British Colonial research—vigorous prosecution of development policy. Daily Gleaner, 16 Sept 1945, p 9

Anonymous (1946a) Mr F.L.B. Evans calls 10 year plan a play with Ministers as comedians. Daily Gleaner, 3 Apr 1946, p 10

Anonymous (1946b) Mining policy for colonies. Daily Gleaner, 10 Dec 1946, pp 1 & 16

Anonymous (1947) Development and welfare in the West Indies. Daily Gleaner, 11 Aug 1947, p 8

Anonymous (1948a) Colonial Office geologist here to survey mineral resources. Daily Gleaner, 5 Jan 1948, p 1

Anonymous (1948b) Jamaica to ask U.K. for Geological Survey grant. Daily Gleaner, 26 Feb 1948, p 8

Anonymous (1948c) Legislature to get final 10-year plan draft soon. Daily Gleaner, 5 Jan 1948, p 1

Anonymous (1948d) Proposed water schemes for island: amendment to firearm law. Daily Gleaner, 6 Mar 1948, p 12

Anonymous (1948e) Members of House irked by delay in answering questions. Daily Gleaner, 10 Mar 1948, p 14

Anonymous (1948f) U.S. Geologists can work in W.I. under ERP. Daily Gleaner, 29 Nov 1948, p 1

Anonymous (1949a) Bringing U.S. geologists into Colonial Service. Daily Gleaner, 10 Jan 1949, p 7

Anonymous (1949b) Institute may get new annex. Daily Gleaner, 6 Feb 1949, p 1

Anonymous (1949c) Building of Institute Annex about to start. Daily Gleaner, 30 Sept 1949, p 12

Anonymous (1949d) Professor V. A. Zans, M.Sc (Photograph). Daily Gleaner, Oct 1949, np

Anonymous (1949e) Dr Zans to make geological survey. Daily Gleaner, 31 Oct 1949, p 8

Anonymous (1950a) Geological Department in process of development. Daily Gleaner, 1 Feb 1950, p 9

Anonymous (1950b) Offices being built for CDC geological surveyors. Daily Gleaner, 9 Jan 1950, p 4

Anonymous (1950c) Extra sum sought to complete geological offices. Daily Gleaner, 18 Aug 1950, p 7

Anonymous (1951a) Geological Survey Dept. now at Institute. Daily Gleaner, 8 Mar 1951, p 9

Anonymous (1951b) Licence given to prospect old hope mine. Daily Gleaner, 9 Jun 1951, p 1 & 11

Anonymous (1952a) Govt. plans to step up geological survey. Daily Gleaner, 21 Nov 1952, p 11

Anonymous (1952b) UN aid sought in geological survey. Daily Gleaner, 22 Nov 1952, p 1

Anonymous (1952c) What about Jamaica's mineral resources. Daily Gleaner, 12 May 1952, p 6

Anonymous (1954a) Considerable achievement' seen in water supplies here. Daily Gleaner, 2 Jan 1954, p 5

Anonymous (1954b) Farewell for B.G. Geological Surveys Head. Daily Gleaner, 2 Jan 1954, p 2

Anonymous (1954c) PNP's plan for progress. Daily Gleaner, 16 Sept 1954, p 10

Anonymous (1955a) Soda ash factory to be set up here. Daily Gleaner, 17 Jun 1955, p 14

Anonymous (1955b) Geologist on oil possibilities here. Daily Gleaner, 26 Mar 1955, p 6

Anonymous (1955c) Oil promise in Negril test. Daily Gleaner, 12 Jul 1955, p 1

Anonymous (1955d) Big oil exploration soon. Daily Gleaner, 26 Oct 1955, p 1 & 14

Anonymous (1955e) History of Stanolind. Daily Gleaner, 26 Oct 1955, p 14

Anonymous (1955f) Geology teaching in school. Daily Gleaner, 23 May 1955, p 8

Anonymous (1955g) Dinner marks formation of local Geologists' Group. Daily Gleaner, 10 Oct 1955, p 11 & 17

Anonymous (1956) Call for geology in school curriculum. Daily Gleaner, 11 Dec 1956, p 17

Anonymous (1957a) IDC man views Canadian mining development. Daily Gleaner, 18 Oct 1957, p 10

Anonymous (1957b) Stanolind now Pan Jamaican Oil Co. Daily Gleaner, 2 Apr 1957, p 1

Anonymous (1957c) West Negril oil search abandoned. Daily Gleaner, 6 Sept 1957, p 1

Anonymous (1957d) Oil search abandoned. Daily Gleaner, 6 Nov 1957, p 16

Anonymous (1957e) Geologists' association holds dinner. Daily Gleaner, 20 Dec 1957, p 14

Anonymous (1958) Awarded Second Doctorate. Daily Gleaner, 27 Feb 1958, p 3

Anonymous (1961) Professor Zans, Director geological survey, dies at 57. Daily Gleaner, 7 Sept 1961, 1

Anonymous (1997a) Aggregates due in days. Daily Gleaner, 18 Jan 1997

Anonymous (1997b) Local contractors unable to provide quality sand—Pickersgill. Daily Gleaner, 1 Jan 1997

Anonymous (1998) Non-skid aggregate being sought by Mining Ministry. Daily Gleaner, 6 Nov 1998

Anonymous (1999) Non-skid aggregate possibility for Jamaica. Daily Gleaner, 11 Aug 1999

Anonymous (2013) A rare find. Jamaica Observer, 24 Oct 2013. https://www.jamaicaobserver.com/news/a-rare-find/. Accessed 23 Mar 2023

Anonymous (2014a) Rare earth project on hold. Radio Jamaica News Online, 31 Aug 2014. http://radiojamaicanewsonline.com/local/rare-earth-project-on-hold. Accessed 23 Mar 2023

Anonymous (2014b) Jamaica now has full ownership of rare earth plant. Radio Jamaica News Online, 10 Oct 2014. http://radiojamaicanewsonline.com/business/jamaica-now-has-full-ownership-of-rare-earth-plant. Accessed 23 Mar 2023

Anonymous (2021) Element US advances $800 million rare earth element project in Louisiana. Louisiana Economic Development News, 3 May 2021. https://www.opportunitylouisiana.gov/led-news/news-releases/news/2021/03/05/elementus-advances-800-million. Accessed 23 Mar 2023

Aratram MR (1999) 140 years of geological assessment in Ja—perspectives from the Mines and Geology Division (MGD). Daily Gleaner, 9 Aug 1999, p B4-B5

Barclay LA, Girvan N (2013) Transnational restructuring and the Jamaican bauxite industry: the swinging pendulum of bargaining power. In: Gendron RS et al (eds) Aluminum ore: the political economy of the global bauxite industry. UBC Press, Vancouver, p 238–267

Barnett S (2020) Opportunities for Ja to capitalise on rare earth elements. The Gleaner, 22 Sept 2020. https://jamaica-gleaner.com/article/commentary/20200922/stevie-barnett-opportunity-ja-capitalise-rare-earth-elements. Accessed 23 Mar 2023

Beard EH (1954) British Guiana. Col Geol Min Res (q Bull Col Geol Surv) 4(1):79

Beard EH (ed) (1957) Jamaica. Col Geol Min Res 6(3):79

Bhalai SS (2007) Landslide susceptibility and zonation, Portland, Jamaica. Mines and Geology Division, Kingston

Bhalai SS (2009) Landslide susceptibility of St. Thomas, Jamaica. Mines and Geology Division, Ministry of Energy and Mining (Jamaica), Scale 1:50,000

Bhalai SS (2010) Landslide susceptibility of Portland, Jamaica: assessment and zonation. Carib J Earth Sci 41:39–54

Bhalai SS (2011) Jamaican Marble: a guide to the varieties of commercial marbles found in Jamaica. Mines and Geology Division, Ministry of Science, Technology, Energy and Mining, Kingston (Bull 12, revised edition)

Bhalai SS (2012) Collaborate to update Jamaica's geological database. Daily Gleaner, 27 Jul 2012. https://jamaica-gleaner.com/gleaner/20120727/cleisure/cleisure4.html. Accessed 10 Jul 2014

Bhalai SS (2018) Geospatial data management strategy for the Mines and Geology Division. Mines and Geology Division, Ministry of Transport and Mining (Jamaica)

Bhalai SS, Harris N (2004) Landslide susceptibility map of St. Mary, Jamaica. Mines and Geology Division,

Ministry of Land and Environment (Jamaica), Scale 1:50,000
Bhalai SS, McKoy O (2020) Economic potential of Jamaican red mud—a desk study. Mines and Geology Division, Ministry of Transport and Mining (Jamaica)
Bhalai SS, Harris N, Green R et al (2007) Landslide susceptibility of Portland, Jamaica. Mines and Geology Division, Ministry of Agriculture and Lands (Jamaica), Scale 1:50,000
Black CV (1976) History of Jamaica. Collins Clear-Type Press, Great Britain
Black CDG, Green GW et al (1972) Preliminary investigation of the Castleton copper prospect. Economic Geology Report of the Geological Survey of Jamaica, vol 4
British Geological Survey (1991) BGS technical cooperation in the Caribbean (1965–90). British Geological Survey, Nottingham
Brookes S (ed) (1984) Jamaica—Geology, scale 1:250,000. Geological Survey Division, Ministry of Mining and Energy and Tourism, Kingston
Brown GF (1982) Memorial to Harlan Richard Bergquist 1908–1982. Geol Soc Am, Memorial Series 14:1–6. https://www.geosociety.org/documents/gsa/memorials/v14/Bergquist-HR.pdf. Accessed 16 Dec 2022
Brown IC, Langner DM (2002) Type and figured specimens in the Geology Museum, University of the West Indies, Mona campus, Jamaica. Geol Curator 7(8):299–304
Bryce R, Harris N et al (1987) Preston Lands—a geological phenomenon. In: Geological Survey Division Biennial Report April 1985–March 1987. Geological Survey Division, Kingston, pp 20–30
Business Information Group (1993) Orvana options Jamaican ground. North Miner, 18 January 1993. http://www.northernminer.com/issuesV2/VerifyLogin.aspx#sthash.4jvDEz7k.dpuf. Accessed 10 Jul 2014
CIDA (1988) Geochemical Survey, Report—Jamaica Metallic Mineral Survey—Phase I. Canadian International Development Agency (CIDA) Project No. 504/0012280, Bondar-Clegg, Ottawa
Chubb LJ (1955a) Letter to the Editor—Geologists' Association. Daily Gleaner, 26 Jul 1955, p 8
Chubb LJ (1955b) A revision of Whitfield's type specimens of the Rudist Mollusks from the Cretaceous of Jamaica, British West Indies. Am Mus Novitates, New York, 15 Mar 1955 (no. 1713)
Chubb LJ (1964) Lucas Barrett, a biography. J Geol Soc Jam 6:38–41
Chubb LJ (1971) Rudists of Jamaica. Palae Am 7(45):161–257
Chubb LJ (1980) The Wall Collection of Jamaican Cretaceous Mollusca in the British Museum (Natural history). J Geol Soc Jam 19:52–53
Chubb LJ, Williams JB et al (1961) Zans Memorial Issue. Geonotes 4(3–4):1–39
Cleevely RJ (1983) World Palaeontological Collections. British Museum (Natural History), London
Comer JB (1974) Genesis of Jamaican bauxite. Econ Geol 69:1251–1264
Conahan FC (1983) Letter to the Hon. Robert W. Edgar, House of representatives, on Peace corps disaster assistance efforts in Jamaica (NSIAD-84-38). US General Accounting Office, 19 Dec 1983. https://www.gao.gov/assets/nsiad-84-38.pdf. Accessed 21 Dec 2022
Cover WA (ed) (1951) The Handbook of Jamaica for the Year 1951. Government Printing Office, Kingston
Cover WA (ed) (1953) The Handbook of Jamaica for the Year 1953. Government Printing Office, Kingston
Crooks R (1987) History and development of geological survey—Jamaica (1800–1983). In: Ahmad R (ed) Proceedings of a workshop on the status of Jamaican geology. Special Publication of the Geological Society of Jamaica, pp 7–13
Davis CE (1973) The mineralogy of Jamaican bauxites. J Geol Soc Jam Proceedings of Bauxite Symposium II, Oct 1973, Special Issue, Pub no. 143:6–20
Davis CE (1989) Jamaica in the world aluminium industry 1938–1973, Volume I. Jamaica Bauxite Institute, Kingston
Davis CE (1995) Jamaica in the world aluminium industry, Volume II 1974–1988, bauxite levy negotiations. Jamaica Bauxite Institute, Kingston
De la Beche HT (1827) Remarks on the geology of Jamaica. Trans Geol Soc Lond 2(2):143–194
Douglas L (1996) Contractor defends stone importation. Daily Gleaner, 19 Dec 1996
Exploration Management Services Pty Ltd (2014) Main Ridge Gold Mine (Jamaica). Exploration Management Services Pty Ltd. http://www.explorems.com.au/projects/main-ridge-gold-mine-jamaica/. Accessed 10 Jul 2014
Fenton AD (ed) (1979) Copper prospects of Jamaica; A geological review. Bull Geol Surv Jam 9
Fenton AD (1981) The Mineral Resources of Jamaica. Bulletin 8, 2nd ed, Geological Survey of Jamaica, Kingston
Franks DM, Pakoun L, Ngonze C (2016) Development minerals: transforming a neglected sector in Africa, the Caribbean and the Pacific. United Nations Development Programme, Brussels
Fulford GD, Lever G, Sato T (1991) Recovery of rare earth elements from Bayer process red mud. United States Patent no. 5030424, 9 Jul 1991 (Assignee: Alcan International Limited)
Garrett RG (1985) Feasibility study mission, September 24 to October 5, 1984. Jamaica metallic mineral survey 221-504-101, Canadian International Development Agency
GSD (1984) Biennial report of the Geological Survey Division from April 1982–March 1984. Geological Survey Division, Ministry of Mining, Energy and Tourism, Kingston
Gleason W (2014) Critical minerals 2014; Rare earths and more discussed at conference. Min Eng 66(10):31

Goldich SS, Berquist HR (1947a) Aluminous lateritic soil of the Sierra de Bahoruco area, Dominican Republic. W. I. U.S. Geological Survey Bulletin No. 953-C:53–84
Goldich SS, Berquist HR (1947b) Aluminous lateritic soil of the Republic of Haiti. W. I. U.S. Geological Survey Bulletin No. 954-C:63–109
Government of Jamaica (1955) Annual report of the Geological Survey Department for the financial year 1953–54. The Government Printer, Kingston
Government of Jamaica (1956) Report on Jamaica for the Year 1955. The Government Printer, Kingston
Government of Jamaica (1957) 1956 Report on Jamaica. Government Printers, Kingston
Government of Jamaica (1958a) Annual Report of the Geological Survey Department, Jamaica for the financial year 1955–1956. The Government Printer, Kingston
Government of Jamaica (1958b) Annual Report of the Geological Survey Department for the year ended 31st March 1957. The Government Printer, Kingston
Government of Jamaica (1961) Annual Report of the Geological Survey Department for the financial year 1959–60. The Government Printer, Kingston
Government of Jamaica (1962) Annual Report of the Geological Survey Department for the year ended 31st March 1961. The Government Printer, Kingston
Government of Jamaica (1963) Annual Report of the Geological Survey Department for the year ended 31st March 1962. The Government Printer, Kingston
Government of Jamaica (1964) Annual Report of the Geological Survey Department for the year ended 31st March 1963. The Government Printer, Kingston
Government of Jamaica (1965) Annual Report of the Geological Survey Department for the year ended 31st March 1964. The Government Printer, Kingston
Government of Jamaica (1966) Annual Report of the Geological Survey Department for the year ended 31st March 1965. The Government Printer, Kingston
Government of Jamaica (1967) Annual Report of the Geological Survey Department for the year ended 31st March 1966. The Government Printer, Kingston
Government of Jamaica (1968) Annual Report of the Geological Survey Department for the year ended 31st March 1967. The Government Printer, Kingston
Government of Jamaica (1987) Geological Survey Division Biennial Report, April 1985–March 1987. Geological Survey Division, Kingston
Government of Jamaica (1995) Geological Survey of Jamaica Report, April 1987–March 1994. Geological Survey Division, Kingston
Government of Jamaica (2015) Disaster Risk Management Act, 2015. Law No. 1—2015, 19 Feb 2015
Government Public Relations Office (1960) The Handbook of Jamaica for 1960. Government Printing Office, Kingston
Government Public Relations Office (1961) The Handbook of Jamaica for 1961. Government Printing Office, Kingston
Greiner H (1965) The Oil and Gas Potential of Jamaica. Bulletin No. 5 of the Geological Survey Department of Jamaica, Geological Survey Department, Kingston
GSD (1987a) Port Antonio, Geological Sheet 28. Geological Survey Division, Ministry of Mining and Energy (Jamaica), Scale 1:50,000
GSD (1987b) Holland Bay, Geological Sheet 30. Geological Survey Division, Ministry of Mining and Energy (Jamaica), Scale 1:50,000
Gustaffson B (1988) Consultant services concerning industrial minerals in Jamaica. Geological Survey Division, Kingston
Harder EC (1949) Stratigraphy and origin of bauxite deposits. Bull Geol Soc Am 60:887–908
Harder EC (1952) Examples of bauxite deposits illustrating variation in origin. In: Problems of Clay and laterite genesis symposium, pp 35–64. A.I.M. & M.E. Inc
Harris N (2002) Rain-induced slope failures and damage assessment in Portland, Jamaica, related to the flood event of 3–4 January 1998. In: Jackson TA (ed) Caribbean geology into the third millennium, transactions of the fifteenth Caribbean Geological Conference. University of the West Indies Press, Kingston, pp 235–255
Hill VG (1971) Preface (Bauxite/alumina symposium 1971). J Geol Soc Jamaica Bauxite/alumina Symp 1971:i
Hill VG (1973) The bauxites of Jamaica. J Geol Soc Jam Proceedings of Bauxite Symposium II, Oct 1973, Special Issue, Pub no. 143:1–5
Hill VG (1978) The clay minerals of Jamaica, or, distribution and potential—clays in Jamaica. Jam J 12(42):64–75
Hill VG (1983) Bauxite. In: LeFond SJ (ed) Industrial minerals and rocks—volume 2. American Institute of Mining, Metallurgical and Petroleum Engineers, Inc. np
Hill VG, Ostojic S (1984) The characteristics and classification of bauxites. In: Jacob Jr L (ed). Proceedings of the 1984 bauxite symposium. SME-AIME, New York, pp 31–48
Hill VG, Sehnke ED (2006) Bauxite. In: Kogel JE, Trivedi NC, Barker JM et al (eds) Industrial minerals & rocks: commodities, markets, and uses. Society for Mining, Metallurgy and Exploration, Inc, Colorado, pp 227–261
Hose HR (1950) The geology and mineral resources of Jamaica. Col Geol Min Res 1(1):11–36
Hughes IG (ed) (1973) The mineral resources of Jamaica. Bulletin 8, Geological Survey Department, Kingston
Hylton HA, Burrell V, Fenton A (1987) Mineral Springs of Jamaica. Bulletin 11, Geological Survey of Jamaica, Kingston

Jackson S (2019) Possible oil reserves upsized for Jamaica. Jamaica Gleaner 27 Feb 2019. https://jamaica-gleaner.com/article/business/20190227/possible-oil-reserves-upsized-jamaica. Accessed 26 Mar 2023

Jackson S (2021) Price tag to drill Colibri oil well in Jamaica: US $35m. The Gleaner, 21 May 2021. https://jamaica-gleaner.com/article/business/20210521/price-tag-drill-colibri-oil-well-jamaica-us35m. Accessed 26 Mar 2023

Jackson S (2022) United Oil still hunting partner for Jamaica exploration. The Gleaner, 27 Apr 2022. https://jamaica-gleaner.com/article/business/20220427/united-oil-still-hunting-partner-jamaica-exploration. Accessed 26 Mar 2023

Jamaica Bureau of Standards (1995) Jamaica National Building Code, Volume 2: Energy efficiency building code, requirements and guidelines, 1994. Jamaica Bureau of Standards, Kingston

Jamaica Information Service (1965) The Handbook of Jamaica for 1965. The Government Printing Office, Kingston

James-Williamson SA (2013) Rudist collections at the University of the West Indies Geology Museum (UWIGM). Caribbean J Earth Sci 45:91–98

Lightbourne RC (1970) Mining. Ministry Paper No. 33/70, File no. 14/1/11. Ministry of Trade and Industry (Jamaica), Kingston, 8 June 1970. http://www.nlj.gov.jm/MinistryPapers/1970/33.pdf. Accessed 20 Mar 2023

Linton L (2019) Decision to subsume PCJ into energy ministry will improve efficiency and oversight. Jamaica Information Service, 18 September 2019. https://jis.gov.jm/decision-to-subsume-pcj-into-energy-ministry-will-improve-efficiency-and-oversight/. Accessed 13 Dec 2022

Lyew-Ayee P (1986) A case for the volcanic origin of Jamaican bauxites. Geol Soc Jam Proc of Bauxite Symp 6:9–39

Martin G (1997) Jamaica was deceived! Pickersgill should never have imported rockstone. Daily Gleaner, 22 Mar 1997

Matley CA (1923) Report on the Work of the Government Geologist and the progress of the geological surveys in Jamaica for economic purposes. Suppl Jamaica Gazette 46(9):225–231

Matley CA (1924a) The geological surveys here. Daily Gleaner, 25 Jul 1924, p 14

Matley CA (1924b) Geological Survey of the Cayman Islands. Daily Gleaner, 20 Jun 1924, p 13

McDonald N (2019) Tullow and Jamaica: why the silence over the big oil and gas find? The Gleaner, 12 May 2019. https://jamaica-gleaner.com/article/focus/20190512/norris-mcdonald-tullow-and-jamaica-why-silence-over-big-oil-and-gas-find. Accessed 26 Mar 2023

McFarlane N (ed) (1977) Jamaica Geology, 1:250,000. Mines and Geology Division, Ministry of Mining and Natural Resources, Kingston

Megginson E (2013) Paulwell opens Jamaica rare earth elements pilot plant. Ministry of Science, Technology, Energy and Mining, Press Release, 23 Oct 2013. http://go-jamaica.com/pressrelease/item.php?id=2562. Accessed 23 March 2023

Meyeroff AA, Krieg A (1977) Petroleum potential of Jamaica. Ministry of Mining and Natural Resources, Kingston

Mines and Geology Division (1975) Annual Report of the Geological Survey Department for the Year ended 31st March 1975. Ministry of Mining and Natural Resources, Kingston

Mitchell SF, James-Williamson SA (2013) Introduction. Carib J Earth Sci 45:1–2

MSET (2019) Petroleum Corporation of Jamaica. Ministry of Science, Energy and Technology, 2 Jul 2019. http://www.mset.gov.jm/2019/07/02/petroleum-corporation-of-jamaica/. Accessed 13 De 2022

Newnham RE (1986) Memorial of George W. Brindley, June 19, 1905-October 23, 1983. Am Miner 71:1058–1059

ODPEM (2019) Who are we? Office of Disaster Preparedness and Emergency Management, Kingston. https://www.odpem.org.jm/the-odpem/. 24 Mar 2023

O'Hara M, Bryce R (1983) A Geotechnical Classification of Jamaican Rocks. Bulletin 10, Geological Survey of Jamaica, Kingston

Orvana Minerals Corporation (2005) Orvana Minerals Corporation Annual Information Form for the Fiscal Year ended September 30, 2005. Orvana Minerals Corporation. http://www.orvana.com/investors/pdf/05-aif.pdf. Accessed 10 Jul 2014

Patterson PJ (1979) Ministry Paper No. 56/79, Reconstruction programme for flood damage in Western Jamaica Consequent on June 12 Flood Rains. Ministry of Foreign Affairs (Jamaica), 11 Dec 1979 (MFA File No. 155/04). https://nlj.gov.jm/MinistryPapers/1979/no.%2056.pdf. Accessed 29 Mar 2023

Paulwell P (2013) Statement to Parliament on Rare Earth Elements. Ministry of Science, Technology, Energy and Mining (Jamaica), 15 Jan 2013. http://www.japarliament.gov.jm/attachments/842_STATEMENT%20TO%20PARLIAMENT%20ON%20RARE%20EARTH%20ELEMENTS.pdf. Accessed 23 Mar 2023

Porter ARD (2017) Jamaican bauxite, a retrospective. iMagiNation Books, Kingston

Porter ARD (2020) Alcan's proposed 1976 bauxite transportation project: a geo-historical account. Anthony Porter (in press)

Richards D, Elliston H et al (ed) (1998) Jamaican Marble. Bulletin 12, Geological Survey of Jamaica (Mines and Geology Division), Kingston

Robinson E (1983) The peat resources of Jamaica and their potential for fuel supply. Petroleum Corporation of Jamaica, Kingston

Robinson E (1996) Charles Alfred Matley: his links with mid-twentieth century geology in Jamaica. Contributions to Geology, UWI, Mona 2:20–26

Robinson E (2012) The Geology Department: the first twenty years, 1961–1980. In: Lyew-Ayee A (ed) Celebrating fifty years of teaching geology at the University of the West Indies—a commemorative booklet. Department of Geography/Geology, University of the West Indies, Kingston, pp 1–8

Robinson E, Gentles H (1980) The Geological Society of Jamaica—its history. J Geol Soc Jam 19:54–59

Robinson E, Gunter G et al (2012) Obituary. Raymond Marcio Wright, Ph.D., CD. 13th June 1940–7th July 2011. Carib J Earth Sci 44:1–3

Rousseau P (1973) Preface (To Proceedings of Bauxite Symposium II). J Geol Soc Jam Proceedings of Bauxite Symposium II, Oct 1973, Special Issue, Pub no. 143:i

Royal Gold Inc (2014) Royalties. Royal Gold Inc. http://www.royalgold.com/business-model/royalties/default.aspx. Accessed 10 Jul 2014

Saunders A (2020) PCJ to be wound up by March 31. Daily Observer 30 Jan 2020. https://www.jamaicaobserver.com/news/pcj-to-be-wound-up-by-march-31/. Accessed 13 Dec 2022

Sawkins JG (1869) Reports on the geology of Jamaica; or Part II of the West Indian Survey, with contributions from G. P. Wall, Lucas Barrett, Arthur Lennox and C. B. Brown, and an Appendix by R. Etheridge, (Memoirs of the Geological Survey, U. K.). Longmans, Green, and Co., London

Schmedeman OC (1948) Caribbean aluminum ores. Eng Min J 149(6):78–82

Schmedeman OC (1950) First Caribbean bauxite development, Reynolds Jamaica Mines Ltd. Eng Min J 151(11):98–100

Scott R (2019) PCJ shutdown—Energy Minister defends impending winding up of state body amid opposition alarm. Daily Gleaner, 14 Sept 2019. https://jamaica-gleaner.com/article/lead-stories/20190914/pcj-shutdown-energy-minister-defends-impending-winding-state-body-amid. Accessed 13 Dec 2022

Sinclair IGL (1967) Bauxite genesis in Jamaica: new evidence from trace element distribution. Econ Geol 62:482–486

Siriunas JM (1993) Draft Report no. 27, Inventory of Metallic Mineral Occurrences in Jamaica. Geological Survey Division, Kingston

Spence C (2021) Large deposits of rare earth elements. Jamaica Information Service, 27 May 2021. https://jis.gov.jm/large-deposits-of-rare-earth-elements/. Accessed 23 Mar 2023

Steers JA (1940) The Cays and the Palisadoes, Port Royal, Jamaica. Geog Rev 30(2):291

Stephenson LW (1942) Preliminary report on Upper Cretaceous Fossils collected by Mr. C. A. Matley in Jamaica, B.W.I, (no publisher listed)

Strickland ERD (1989) De nominibus quaestor. J Geol Soc Jam 26:28–32

Sugita K, Ota Y, Taguchi Y et al (2014) Rare earth element recovery method. US Patent Application Pub US2014/0283652 A1, 25 Sept 2014. (Assignee: Nippon Light Metal Company, Ltd, Tokyo, Japan)

Sugita K, Kobayashi Y, Taguchi Y et al (2016) Method of recovering rare-earth elements. US Patent 9,228,248 B2, 5 Jan 2016, (Assignee: Nippon Light Metal Company, Ltd, Tokyo, Japan)

Tien JC (2013) China's rare earth minerals: reserves, supply and demand. Min Eng 65(12):44–49

Trechmann ST (1922) The *Barrettia* beds of Jamaica. Geol Mag 59:501–514

Trechmann CT (1924) The Cretaceous limestones of Jamaica and their mollusca. Geol Mag 61:385–410

Tyson CF (1983) Preliminary environmental impact statement on the proposed Montego Bay, Jamaica Quarry Zone. Mines and Quarries Division, Kingston

United Nations (1988) Industrial minerals development, Jamaica—Project findings and recommendations (DP/UN/JAM-82-008/1). United Nations Department of Technical Co-operation for Development (Executing agency for United Nations Development Programme), New York

United Oil and Gas (2023) Jamaica. United Oil and Gas. https://www.uogplc.com/operations/jamaica/. Accessed 26 Mar 2023

Van Zandt FK (1976) Boundaries of the United States and the several states. Geological Survey Professional Paper 909, United States Geological Survey, United States Government Printing Office

Versey HR, Prescott GC (1958) Progress report on the geology and groundwater resources of the Clarendon Plains, Jamaica, W.I. Occasional Paper No. 1, Geological Survey Department (Jamaica), The Government Printers, Kingston

Wagh AS, Pinnock WR (1987) Occurrence of scandium and rare earth elements in Jamaican bauxite waste. Econ Geol 82:757–761

Walker HS (1971) Address on the bauxite/alumina industry of Jamaica. J Geol Soc Jam Bauxite/alumina Symp 1971:1–2

Water Resources Authority (2011) History, How it all began. Water Resources Authority (Jamaica). http://www.wra.gov.jm/dynaweb.dti?dynasection=history&dynapage=resource_history. Accessed 30 Nov 2011

Whitfield RP (1897a) Descriptions of species of Rudistae from the Cretaceous rocks of Jamaica, W.I., collected and presented by Mr F. C. Nicholas. Bull Amer Mus Nat Hist, 9:185–196, (plates 6–22)

Whitfield RP (1897b) Observations on the genus *Barrettia* Woodward, with descriptions of two new species. Bull Amer Mus Nat Hist 9:233–246

Wigton (2023) Wigton Wind Farm Limited. Wigton Wind Farm Limited. https://wwfja.com/wigton-wind-farm/. Accessed 26 Mar 2023

Williams JB (ed) (1962) Geology at the University of the West Indies. J Geol Soc Jam (Geonotes) 5(1 and 2):29–30

Williams P (2006) Elusive gold. Jamaica Observer, 1 Jan 2006. http://www.jamaicaobserver.com/rows/95739_Elusive-gold. Accessed 10 Jul 2014

Wilson HE (1985) Down to Earth: one hundred and fifty years of British Geological Survey. Scottish Academic Press Ltd., Edinburgh

Woodward SP (1862) Some account of *Barrettia*, a new and remarkable fossil shell from the Hippurite limestone of Jamaica. Geologist 5:372–377

Wright RM (2010) Energy, The new agenda. Raymond M. Wright, Kingston

Zans VA (1951a) Economic geology and mineral resources of Jamaica (Bulletin No. 1 of the Geological Survey Department of Jamaica). Geological Survey Department, Kingston

Zans VA (1951b) The search for oil in Jamaica. Daily Gleaner, 2 Nov 1951, p 8

Zans VA (1952a) Provisional Structural and Geological Map of Jamaica (Scale 1 inch to 2.698 miles). Geological Survey Department, Kingston

Zans VA (1952b) Bauxite resources of Jamaica and their development. Col Geol Min Res 3(4):307–333

Zans VA (1953) Geological Map of Jamaica (1 inch to 10 miles). In: Cover WA (ed) The Handbook of Jamaica for the Year 1953. Government Printing Office, Kingston, np

Zans VA (1955) Geology and mining in Jamaica (Jamaica 300 Supplement). Daily Gleaner, 10 May 1955, p 19

Zans VA (1958) Jamaica Geology (Kingston: Geological Survey Department, 1958, 2200/10/59/5993/O.S.), scale 1:250,000. In: Zans VA, Chubb LJ, Versey HR, et al (1963) Synopsis of the geology of Jamaica; An explanation of the 1958 provisional geological map of Jamaica (Bulletin No. 4 of the Geological Survey Department of Jamaica). Geological Survey Department, Kingston

Zans VA (1959) Recent views on the origin of bauxite. Geonotes 1:123–132

Zans VA, Chubb LJ, Versey HR et al (1963) Synopsis of the Geology of Jamaica; An Explanation of the 1958 Provisional Geological Map of Jamaica (Bulletin No. 4, Geological Survey Department of Jamaica). Geological Survey Department, Kingston

6 Transforming Jamaica's Mining Sector

Abstract

The journey of the development of Jamaica's Mining Sector is instructive for charting the way forward. A handful of simple principles are described, which if inculcated, offers significant benefits towards true transformation of the Sector as it continues. These are (1) a positive mindset towards minerals; (2) collaboration in knowledge; (3) demonstration of responsibility towards minerals development; (4) equity for all stakeholders; and (5) accountability among stakeholders. These principles can be truly successful when the country, company, and the community have consensus.

6.1 Guiding Principles for Transformation

The history of mining and geological surveys in Jamaica characterizes a journey of a small country managing its geological assets. Recorded are the successes, the struggles, the accomplishments, and the failures. It also emphasizes several pertinent principles which, in the opinion of the author, should be inculcated for Jamaica's Mining Sector to fundamentally evolve into a vibrant, high performing, and profitable entity in the economy. The list is not exhaustive, but some of the key values are (1) a positive mindset towards minerals; (2) collaboration in knowledge; (3) demonstration of responsibility towards minerals development; (4) equity for all stakeholders; and (5) accountability among stakeholders. In examining these, the reader will recognize three important stakeholders: (1) the country; (2) the mining company; and (3) the communities in which they operate. All three are interdependent and essentially must agree for a mining venture to be remunerative for all. The country and community rely on the expertise of the company to monetize their natural resources, while the company, to be profitable, will require the relevant permissive or possessive rights, along with the social licence to operate (Raufflet et al. 2013).

6.1.1 A Positive Mindset to Minerals

When the first commercial shipment of bauxite was being loaded in Ocho Rios in 1952, the headlines of the *Daily Gleaner* stated, 'Red Gold going for first time' (Anonymous 1952). Minerals are but one of the natural resource assets for a country, and it is without question that nations seek the optimum benefit. Unfortunately, in recent years in Jamaica, the intrinsic worth of minerals has been negated. Bauxite mining, which is the chief activity in the Mining Sector, has been the target of heavy criticism. Various groups have alleged loss of their

S. Bhalai, *The History of Mining and Geological Surveys in Jamaica*, Historical Geography and Geosciences,
https://doi.org/10.1007/978-3-031-42604-9_6

livelihoods and have complained about environmental destruction, impacts on health, deforestation, dust, emissions and water pollution, improper behaviour from company officials, unfulfilled promises of reclamation and provision of land titles when residents are relocated, among other negative impacts (Haye 2022; McCaulay 2022; Virtue 2022; Anonymous 2023; Neita 2023).

There is hardly mention of the millions of dollars expended by the companies on environmental controls, dust mitigation measures, land rehabilitation practices, forestry development (such as the 110,000 trees planted by one company in 2022), and protection of cultural assets, along with making the relevant payouts for compensation as prescribed by legislation. Many fail to recall the over 70 years of gainful employment provided to Jamaicans and the improved way of life provided to the operating communities and the country as a whole, resulting from the income from bauxite mining. There is failure to mention improvements in agriculture (dairy schemes, castor bean cultivation, and the Bauxite Community Development Programme implemented by the Jamaica Bauxite Institute). In addition, the bauxite and alumina industry has supported national budgets (from the Capital Development Fund), community development, education (schools, scholarships such as the Glencore and Rio Tinto Scholarships, training, and research support), housing (including developments on rehabilitated lands), disaster assistance (e.g. during the June 1979 floods), development of five major ports, roads (such as the Spur Tree and Melrose bypasses), water supplies, among many other benefits (Patterson 1979; Davis 1995; JBI 2021; Gilchrist 2022; Neita 2022).

In the first 30 years of bauxite mining there was no need to convince Jamaicans of the benefits of the industry. In 1970, the Mining Sector had contributed approximately 16% to the country's Gross Domestic Product (GDP) and was considered to have the greatest potential for growth (Walker 1971). However, in 2021, the Sector had fallen to 1.5% of GDP (PIOJ 2021). Despite global conditions, it is still possible to restore the Sector to gainful contribution. However, it demands firstly, a shift in national mindset. Minerals are valuable tangible assets and not just stones. In the case of bauxite, Jamaica should again see it as the respected 'red gold' versus the derogatory 'red dirt'.

6.1.2 Collaboration in Knowledge

Compared to several decades ago, Jamaica now abounds in research capacity in both mining and geology. These reside in the knowledge centres of Government, academic institutions, and companies. Unfortunately, the knowledge oftentimes become contained, stagnant, and rarely shared. Nearly all the accounts of knowledge sharing throughout the history of mining and geological research in Jamaica have stated successful outcomes. The bauxite companies and the Government worked together in the 1940s and 1950s to develop an industry which drove Jamaica's economic advancement. By combined efforts the large quantities of subeconomic material were rapidly developed to establish a prosperous mining industry (Zans 1952). Through symposia on bauxite, industrial minerals, and geology in the 1970s and 1980s, industry technocrats converged to share their experiences and best practices. Oftentimes this enabled companies to understand their challenges, adopt new techniques to make their processes efficient, and resulted in their operations being profitable (e.g. Hill 1971; Lyew-Ayee 1981; Ahmad 1987). Another worthwhile example is significant contributions of faculty and students of the Geology Department at the University of the West Indies in the 1970s, where their work on Jamaican stratigraphy and structure were included by the Geological Survey in the national geological maps (Bhalai 2012). Opportunities for similar collaborations arise continually, whether for updating the national geological maps (which serve as the basis for mineral discoveries) or conducting applied research towards development of mineral-based products to diversify the mineral portfolio of the country and add value which can bring greater economic gains.

6.1.3 Demonstrate Responsibility for Mineral Development

Jamaica had been the beneficiary of many programmes providing funding and technical assistance through multilateral organizations and bilateral cooperation from agencies and countries in North America and Europe. In recent times there has been interventions from the United Nations Development Programme (UNDP) (e.g. Lewis et al. 2017), the Inter-Governmental Forum for Mining, Minerals, Metals and Sustainable Development (IGF) (e.g. IGF 2019; Crawford et al. 2020, and other support on policy and fiscal regimes), and the Inter-American Development Bank (IDB). During the periods of these intervention, there is usually a peak in activities and a dedication in effort towards developing geological sciences and minerals. Activities and efforts seemingly trickle when these programmes end.

The country, as a way forward, should consider advancing its own development agenda for geological surveys and mining without awaiting the impetus and resources brought about through international funding and technical support. The country may even consider reducing the dependency and reliance on the sustenance provided by these international programmes and consider investing its own resources in its own development projects. By exercising self-sufficiency, the support from the international entities can be deployed to programmes in less fortunate countries where the need may be greater. Jamaica's technical repositories and knowledge centres such as the Tertiary institutions (e.g. the University of the West Indies, University of Technology, and others) and Government research entities (e.g. Mines and Geology Division, Jamaica Bauxite Institute, and the Scientific Research Council) should have active advisory and implementation roles in the development of the geological sciences and the Mining Sector. Knowledge products from various international organizations such as the IDB, IGF, Natural Resource Governance Institute, and the World Bank, along with those from technical research entities like the Jawaharlal Nehru Aluminium Research Development and Design Centre (India) and The Sustainable Minerals Institute (Australia) are useful references to remain current and guide in industry best practices.

The Jamaican Government had taken the bold step to formally approve the implementation of its National Minerals Policy (Government of Jamaica 2020) which addresses the challenges faced by the Mining Sector for decades. This policy started out in 2000 as a policy to guide quarry operations, but it was recognized that a broader policy addressing the entire Mining Sector was in fact needed. After two decades of consultations the document was finally accepted for implementation. The challenges are listed as policy issues and are addressed through approximately 190 actions for execution. This includes establishing a new organization that will lead Government research programme in the geosciences with mineral development as a key function.

6.1.4 Equity for All Stakeholders

Davis (1995) using the various bauxite negotiation periods in Jamaica, appropriately presented the selfishness of human tendency to seek advantage. In the context of mining agreements between Government and the mining companies the need for equity in relationships is highlighted. The country and communities should be conscious that companies must make fair returns on their investments of capital and technology. Further, they should recognize that companies compete in a global marketplace which has bearing on the productivity and profitability of their operations.

On the other hand, mining companies should be cognizant that the country seeks a fair value for the use of their raw material resources. Natural resource income is used to adequately provide the infrastructure and services that enable the same companies to operate, and which also sustain the commerce that the employees of the companies depend, oftentimes in the communities hosting the mining operations. Companies should be mindful that citizens, which, in the case of Jamaica, are oftentimes in rural areas, frequently must endure the disturbances, disruptions and sometimes dislocations associated with their mining activities.

With the various interests in mind, consideration and thoughtfulness is a requirement for all stakeholders, and it is not beyond any group, for these to be reasonable expectations. Minerals can be amply remunerative, and all parties can be satisfied if equity is the basis of the relationships.

6.1.5 Accountability Among Stakeholders

Research has demonstrated that stronger resource-governanced countries have better educated, longer-lived, and wealthier citizens (Kaimal 2021). Government is accountable to their people for the management of the country's mineral resources. Ultimately, the extractive industry should be seen as a driver for achieve social inclusion, environmental sustainability, and economic development. Considering this, relevant mechanisms should be in place to ensure that mining companies adhere to the rules of business as stated in the terms of their agreement with the country. Coupled with strong laws should be stronger enforcement. Too often the Government may be timid, carefree, or in a state of 'official inertia' where situations and breaches are left to degrade, and company debts become compounded. The responsibilities of mining companies, oftentimes environmental and fiscal, are neglected for prolonged periods until the stakeholder relationships deteriorate. Fortunately, non-governmental organizations (NGOs) provide the stimulus of concern, especially about environmental management, and will seek to illuminate with the 'sunshine of transparency'. On the other hand, fiscal matters such as debts for royalties and taxes are generally not reported and more often are left to accrue for many years, without action or remedy from what appears as a lethargic Government, including its Civil Service. This breakdown may end in extreme measures being applied, such as revocation of a mining right. However, it is detrimental to a country to revoke the mining rights of a company, as the decision to permit mining oftentimes conforms to the greater plan for positive territory transformation. However, it is much easier for companies to shutter business and not pay their debts. Creative accounting measures enable companies to cite losses as reasons (or rather, as excuses) for not paying a country.

The 2011 Guinean Mining Code (GMC) may serve as a model for the review and revision of Jamaica's mining laws which lays out the rules for mineral operations. The 2011 GMC repealed the 1995 mining code and represents a transformation of the law. Bhatt (2013) aptly describes the new law:

> The 2011 [Guinean Mining Code] GMC was ground-breaking in a number of areas. It was designed to ensure that the government obtains a fair share of the country's natural resource wealth, with the intention that capital can be ploughed back into development policies such as better healthcare and housing and improved infrastructure. Laudable principles of good governance, transparency, local development and employment, redistribution of funds and environmental and social issues feature strongly in the GMC. Clearly the GMC was drafted with the overarching goal that increased government revenue applied in accordance with these principles would create the perfect set of conditioning circumstances to enable the country to lift itself out of poverty and prevent any potential conflict associated with natural resources. (Bhatt 2013)

Undoubtedly, the new GMC was met with criticism from the mining companies which cited that the increased taxation provisions, for example, would erode profitability and discourage investment in Guinea's mines. This resulted in the law being suspended until 2013 when a law amending the GMC was promulgated. The 2013 amendments to the mining code maintained the principles of the 2011 GMC where it endeavoured 'to strike a fair balance between providing a stable and attractive environment for investors and increasing the environmental and social benefits from mining for local communities' (Bhatt 2013).

Government therefore has the responsibility for having better rules, vigilant refereeing, and creating a level playing field in mineral governance. Governance must be implemented. Strong rules and regulations favour self-regulation and leads to better standards. Proper enforcement of the rules requires Government to step forward and address corruption and conflicts. Standardizing the terms of operation for mining companies communicates transparency in the Sector. Free riders pose a danger through incompetence and promote corruption. This includes the elimination of multiple and individually tailored fiscal regimes, and customized terms as recommended in Crawford et al. (2020). Government should have their own governance machinery in place to ensure that companies conform to the terms of their agreement including the timely payment of all remunerations for minerals. Participation in a voluntary governance standard such as the Extractive Industries Transparency Initiative is not a substitute for the active, vibrant and sustained operationalization of a Government's own robust accountability mechanism.

National Hero Norman Manley once said that Jamaica does not lack ideas; what we really lack is fixity of purpose and continuity of effort. A prerequisite for fixity of purpose and sustaining effort is an acceptable level of agreement and unity around ideas (Duncan 2002). Importantly too are the words of Lewis Carroll who reminds us that if we don't know where we are going, any road will get us there. Jamaica's journey continues.

References

Ahmad R (1987) Preface (The status of Jamaican geology). Geol Soc Jamaica, Proc Workshop Status Jamaican Geol 10:v

Anonymous (1952) 'Red Gold' going for first time, bauxite ore loading starts at Ocho Rios. The Daily Gleaner, 28 May 1952, p 1

Anonymous (2023) Tax threat, Gov't warns 'high stakes' bauxite shutdown risks $6b in new taxes. The Gleaner, 27 Mar 2023. https://jamaica-gleaner.com/article/lead-stories/20230327/tax-threat. Accessed 27 Mar 2023

Bhalai SS (2012) Collaborate to update Jamaica's geological database. Daily Gleaner, 27 Jul 2012. https://jamaica-gleaner.com/gleaner/20120727/cleisure/cleisure4.html. Accessed 10 Jul 2014

Bhatt K (2013) The 2011 Guinean Mining Code: reducing risks and promoting social benefits in Africa. S Afr J Intl Aff 20(2):247–270

Crawford A, Bliss M, Davis N (2020) IGF Mining Policy Framework Assessment: Jamaica. International Institute for Sustainable Development, Canada

Davis CE (1995) Jamaica in the world aluminium industry, Volume II 1974–1988, Bauxite levy negotiations. Jamaica Bauxite Institute, Kingston

Duncan DK (2002) Wanted, a fixity of purpose. Jamaica Gleaner, 26 Mar 2002. http://old.jamaica-gleaner.com/gleaner/20020326/cleisure/cleisure3.html. Accessed 28 Mar 2023

Gilchrist C (2022) Discovery Bauxite plants 100,000 trees. The Gleaner, 19 Dec 2022. https://jamaica-gleaner.com/article/news/20221219/discovery-bauxite-plants-100000-trees. Accessed 28 Mar 2023

Government of Jamaica (2020) The National Minerals Policy—fostering sustainability in Jamaica's minerals sector. Ministry of Transport and Mining, Kingston

Haye P (2022) Environmental impact of mining. https://jamaica-gleaner.com/article/letters/20220810/environmental-impact-mining. Accessed 25 Mar 2023

Hill VG (1971) Preface (bauxite/alumina symposium 1971). J Geol Soc Jam Bauxite/alumina Symposium 1971:i

IGF (2019) Mineral valuation and sustainability: Jamaica's bauxite industry. Intergovernmental forum on mining minerals, metals and sustainable development, Jamaica workshop 22–24 July 2019. https://www.igfmining.org/beps/news/mineral-valuation-and-sustainability-jamaicas-bauxite-industry/. Accessed 28 Mar 2023

JBI (2021) Bauxite Community Development Programme. Jamaica Bauxite Institute, Kingston. https://jbi.org.jm/bcdp/. Accessed 24 Mar 2023

Kaimal S (2021) New thinking, different skills, unusual collaborations for the future. In: Virtual Annual General Meeting, IGF 2021, 19 Oct 2021

Lewis D, Gordon P, Guemás M et al (2017) Baseline Assessment of Development Minerals of Jamaica. ACP-EU Development Minerals Programme, implemented in partnership with United Nations Development Programme, United Nations Development Programme, Kingston

Lyew-Ayee A (1981) Preface (Industrial Minerals Symposium 1981). J Geol Soc Jamaica Proc Indus Min Symp September 1981:iii

McCaulay D (2022) On choosing not to know: the bauxite-alumina industry in Jamaica. The Gleaner. https://jamaica-gleaner.com/article/focus/20220313/diana-mccaulay-choosing-not-know-bauxite-alumina-industry-jamaica. Accessed 27 Mar 2023

Neita L (2022) Tell the whole story. Jamaica observer, 9 January 2022. https://www.jamaicaobserver.com/columns/tell-the-whole-story-of-bauxite/. Accessed 30 Oct 2022

Neita L (2023) An injunction not only on mining but on huge chunk of economy. The Gleaner, 12 Mar 2023. https://jamaica-gleaner.com/article/focus/20230312/lance-neita-injunction-not-only-mining-huge-chunk-economy. Accessed 28 Mar 2023

Patterson PJ (1979) Ministry Paper No. 56/79, Reconstruction programme for flood damage in Western Jamaica consequent on June 12 flood rains. Ministry of Foreign Affairs (Jamaica), 11 December 1979 (MFA File No. 155/04). https://nlj.gov.jm/MinistryPapers/1979/no.%2056.pdf. Accessed 29 Mar 2023

PIOJ (2021) Economic and Social Survey of Jamaica, 2021. Planning Institute of Jamaica, Kingston

Raufflet E, Baba S, Perras C et al (2013) Social license. In: Idowu SO, Capaldi N, Zu L, Gupta AD (eds) Encyclopedia of corporate social responsibility. Springer, Berlin, Heidelberg. https://doi.org/10.1007/978-3-642-28036-8_77

Virtue E (2022) IACHR serves 20-day ultimatum on Gov't over bauxite mining dangers. The Gleaner, 19 Dec 2022. https://jamaica-gleaner.com/article/lead-stories/20221219/iachr-serves-20-day-ultimatum-govt-over-bauxite-mining-dangers. Accessed 28 Mar 2023

Walker HS (1971) Address on the bauxite/alumina industry of Jamaica. J Geol Soc Jam Bauxite/Alumina Symposium 1971:1–2

Zans VA (1952) Bauxite Resources of Jamaica and their Development. Col Geol Min Res 3(4):307–333

Index

S. Bhalai, *The History of Mining and Geological Surveys in Jamaica*, Historical Geography and Geosciences, https://doi.org/10.1007/978-3-031-42604-9